Advances in Adaptive Ultrasonics

Providing an overview of a new generation of ultrasonic technology, *Advances in Adaptive Ultrasonics* explores how ultrasonic devices can harness the properties of advanced materials, including shape memory alloys and metamaterials.

The applications of ultrasonic devices range from surgery, drilling, and welding to sonar and energy harvesting. This book demonstrates how engineers can overcome common issues within the field of ultrasonics, such as precision control and choice of materials. Beginning with an overview of ultrasonic technology as it is currently understood, the book goes on to discuss the newest iteration in the form of adaptive ultrasonics and the benefits this can offer to multiple industries. Key topics include advanced materials, notably phase-transforming shape memory alloys, and the principles of adaptive ultrasonic transducer design. The book also covers measurement approaches for characterising adaptive ultrasonic devices and provides an overview of potential applications for the technology.

This book will be of interest to students and engineers in the field of ultrasonic surgery, industrial engineering, welding, and mechanical materials.

Advances in Adaptive Ultrasonics

Andrew Feeney

CRC Press
Taylor & Francis Group
Boca Raton London New York

CRC Press is an imprint of the
Taylor & Francis Group, an **informa** business

Designed cover image: Shutterstock

First edition published 2025
by CRC Press
2385 NW Executive Center Drive, Suite 320, Boca Raton FL 33431

and by CRC Press
4 Park Square, Milton Park, Abingdon, Oxon, OX14 4RN

CRC Press is an imprint of Taylor & Francis Group, LLC

© 2025 Andrew Feeney

ISBN: 978-1-032-34845-2 (hbk)
ISBN: 978-1-032-34847-6 (pbk)
ISBN: 978-1-003-32412-6 (ebk)

DOI: 10.1201/9781003324126

Typeset in Times
by SPi Technologies India Pvt Ltd (Straive)

Contents

Preface

This book focuses on some of the key advances in adaptive ultrasonics, a specialist area of research and development centred on the changing properties of an ultrasonic device or process in response to external stimuli, towards intelligent and multifunctional systems. The content is intended for all those interested, from graduates through to professional practitioners, as a primer which will hopefully be instructive for all those exploring concepts associated with adaptive ultrasonics.

The idea for writing this book originated from two places. The first was my academic research, where we have recently made significant steps forward in understanding how we can integrate fascinating advanced materials into traditional configurations of ultrasonic devices. Nitinol, a binary shape memory alloy of nickel and titanium, is the core material central to the mechanical and materials engineering perspectives of adaptive ultrasonics as described in this book, and it is recognised that this material serves as a valuable foundation for other researchers and engineers to move forward into new technology spaces. I believe there are a multitude of possibilities, but it will likely take many years of research and development to fully understand how to unlock the full potential of Nitinol and other shape memory materials. Of course, the content of this book is not limited to shape memory materials, but also metamaterials, with significant implementation potential, provided we can overcome the manufacturing challenges. I am in no doubt that this will happen in the next few years. On the other side of all this, we have recently seen significant developments in signal processing and artificial intelligence capabilities implemented in the field of ultrasonics, and we must continue to be conscious of how these can be responsibly used to enhance our ultrasonic technologies and the impactful outcomes we are all seeking. Adaptive ultrasonics is a bridge between materials science, dynamics, and computational engineering, and therefore has the potential to be a far broader discipline than it is now. It will take major progress in each of these areas to realise a new generation of ultrasonic device for industry and medicine, and in all aspects of these.

The second source of inspiration for this book was to respond to the changing nature of the research and development environment. Working practices around the world have evolved, from remote working, advanced cloud-based computing, advanced fabrication techniques, and additive manufacturing, to the responsibilities and expectations on scientists and engineers in both academia and industry alike. Many advanced materials are expensive, complicated to manufacture, or difficult to obtain. It has been a concern that these are major barriers to new innovations in science and engineering, and so one of the purposes of this book is to open as many new ways as possible for interesting and impactful research and development, with pathways and innovation directions in terms of computing and simulation, as well as the complex manufacturing solutions required. A perspective and assessment of modern practice is included in this book, with projections for future development and what realistically needs to happen for our technology base to progress as we may require. It also aims to give some useful advice for earlier career scientists and engineers, primarily relating to the areas of research requiring particular attention, and strategies for engaging in this field, given the scarcity of material and generally high expense.

Acknowledgements

There are many people I would like to thank for making this book happen. I have been fortunate in my career to have had many influential, knowledgeable, and supportive mentors and colleagues, at each stage of my career. They are too numerous to mention, but the first I must publicly make known is my doctoral adviser and current colleague, Professor Margaret Lucas, Professor of Ultrasonics and the Regius Chair of Civil Engineering and Mechanics at the University of Glasgow. I am indebted to her for passing on her knowledge and wisdom through the years, and I am confident I am a better academic and engineer as a result. I would also like to thank Dr Andrew Mathieson, who has supported me since my first day as a doctoral student. Without his deep knowledge of ultrasonic transducers and experimental mechanics, I would not have been able to make it this far. I would also like to give special thanks to Peter McKenna and Ewan Bremner of the University of Glasgow's James Watt School of Engineering, for their specialist expertise in graphics and illustrations, and to Alastair Feeney, for valuable proofreading and advice.

A major turning point in my career was when I took up a position as Research Fellow at the University of Warwick between 2016 and 2020. Professor Steve Dixon, Director of the Centre for Industrial Ultrasonics in the Department of Physics, became both a friend and a colleague, and I have been fortunate to learn from him and enjoy his support ever since. Similarly, Dr Lei Kang kindly educated me on what it takes to be a truly excellent experimental researcher, and I also thank him for the support I have received over many years.

I would like to thank everyone who has been part of my *Adaptus* research group since its inception in 2020. Each person has contributed in a meaningful way to where the discipline of adaptive ultrasonics is positioned today, and I have been proud to work with each of them. I only hope they have had a similar experience.

Little in my career has been achievable without the support of my family, and I must sincerely thank them for all the support I have received. I want to make a special mention of my parents and siblings, who have always been highly supportive of me in each of my endeavours, and they are always glad to hear of the progress I have been able to make with my colleagues and my research group. It has made the difficult days worthwhile. Most importantly, I want to thank my own family, and without their support and the necessary distraction they give me in challenging times, none of this would have been possible. Thank you!

I hope that the material in this book provides a useful starting point for scientists and engineers with interests in ultrasonics and the innovation of novel concepts for the future. I have attempted to aim this material as broadly as possible, and I especially hope that those in earlier career stages and those who are relatively new to the discipline can get the most out of it and to have success in their own areas of ultrasonics. For as long as I am a practicing academic, I invite anyone with interests in this material and with advice or ideas they wish to discuss, to freely contact me and I will gladly engage in conversation.

Andrew Feeney
University of Glasgow
June 2024

Author

Andrew Feeney, PhD, earned a master's in mechanical engineering with honours of the first class at the University of Glasgow in the United Kingdom in 2010 and a PhD in mechanical engineering in 2014 at the same institution, investigating the incorporation of the shape memory alloy Nitinol in flextensional cymbal transducers for the eventual development of tuneable ultrasonic devices. This research laid the groundwork for this book.

Since then, Dr Feeney has undertaken several cross-disciplinary postdoctoral research projects in a range of application areas including ultrasonic surgery and subsea exploration. In 2016, he joined the Centre for Industrial Ultrasonics as Research Fellow in the Department of Physics at the University of Warwick, leading a new area of research activity in high-frequency flexural ultrasonic transducers. During this time, he collaborated with several industry partners to engineer novel configurations of the flexural ultrasonic transducer for hostile environments, including those of elevated pressure and temperature. This research significantly expanded the practical application of the flexural ultrasonic transducer beyond their traditional uses in automotive systems.

In 2020, Dr Feeney joined the Centre for Medical and Industrial Ultrasonics as Lecturer/Assistant Professor in the James Watt School of Engineering at the University of Glasgow before being promoted to Senior Lecturer/Associate Professor in 2023. Dr Feeney leads the Adaptive Ultrasonics (Adaptus) Research Group, which tackles key challenges in engineering adaptive ultrasonic devices and systems for a new generation of intelligent, multifunctional, and tuneable medical and industrial technologies. The Adaptus Research Group undertakes fundamental research on advanced materials, principally those which exhibit shape memory behaviour, and metamaterials. Dr Feeney's research group focuses on the integration of these materials with a broad range of acoustic and ultrasonic devices and electromechanical systems in order to tailor and optimise dynamic performance. The Adaptus Research Group is centred around four principal themes: shape memory alloys and metamaterials, medical devices, industrial sensors and actuators, and sustainable manufacturing. Dr Feeney's research group includes several postdoctoral researchers and doctoral students, and his research is funded by grants from a combination of UK Research and Innovation, European Research Council, and industry in the order of millions GBP.

At the time of publication, Dr Feeney is a Chartered Engineer and Member of the Institution of Mechanical Engineers (IMechE) and a Member of the Institution of Electrical and Electronics Engineers (IEEE). He has published over 50 peer-reviewed journal and conference articles, is a regular peer reviewer of the major journals in the fields of ultrasonics, ferroelectrics, mechatronics and interdisciplinary science, and is a reviewer of grant proposals for UK Research and Innovation, principally the Engineering and Physical Sciences Research Council.

1 A Brief Overview of Ultrasonics

1.1 MOTIVATION

The field of ultrasonics is fast growing on an international scale, where we are now well equipped as a society to deliver a wide range of intricate and high-performance industrial and medical procedures, both in terms of measurement and for power ultrasonics, or often referred to as destructive, applications. There have been notable advances in medical imaging, nondestructive testing and evaluation for composites and advanced compositions of alloy, ultrasonically assisted surgery, welding and wire bonding, and flow measurement in hostile environments in the last few decades. The applications for which ultrasonic technologies have been transformative in recent years are numerous, but one key feature which links them all together is that they have been made possible because of significant advances in material selection and control, combined with sophisticated manufacturing techniques. These include a broad array of additive manufacturing capabilities, of both metallic and non-metallic materials, which have enabled the fabrication and development of advanced materials with properties and geometries previously not possible.

As advanced manufacturing capabilities have developed, there has been growing focus on the design and fabrication of acoustic and ultrasonic devices incorporating dynamically tuneable and multi-frequency characteristics. Many concepts are only practically possible through advanced manufacturing strategies, and it is only in recent years that they have moved from the theoretical or simulation domains towards prototyping and experimental investigation. The quality of a device being tuneable, or in the context of this book, *adaptive*, is not a new principle, as the background material in this book will demonstrate. However, it is evident that a new branch of ultrasonic technology is emerging, which encompasses the integration of advanced, smart, multifunctional, and adaptable materials in systems and devices. Adaptive features are made possible through advances in manufacturing capabilities, and they are required in response to changing technology demands. The Cambridge English Dictionary defines the term *adaptive* in the contexts of machinery, mechanisms, and mechanical processes as (Cambridge Dictionary Online):

'having an ability to change to suit changing conditions'.

This is a fundamental tenet of this book. The field of ultrasonics has influenced virtually all areas of society for many decades, in applications as varied as surgery,

DOI: 10.1201/9781003324126-1

food cutting, and welding, to car parking sensor systems, haptic perception, and underwater communications. In general, transducers for applications requiring relatively low ultrasonic frequencies (approximately 20–100 kHz), such as the Langevin configuration for ultrasonically enhanced orthopaedic surgery, tend to be tuned to a specific mode of vibration, for example, that of a cutting tip. The dynamic response of such a device is often tailored and defined for a specific application, where typically the concept of adaptive behaviour is not integrated to the development process of the transducer. However, by innovating a new generation of ultrasonic technologies able to adapt and respond to their environments, we can instigate the development of a raft of new possibilities to transform conventional ultrasonic processes into potentially more efficient and high-performance systems for a wider range of applications.

For example, using an adaptive approach, it may be possible to engineer an ultrasonic transducer that is able to undertake faster or safer ultrasonic surgery, resulting in lower tissue damage and expediting patient recovery. An adaptive ultrasonic device could be designed to optimally interact with soft or hard tissue, depending on the physical properties of the device which influence resonance, that are actively controlled. We can also consider an adaptive ultrasonic device which can be remotely repaired, or healed, through a controlled stimulus, removing the need for intrusive, expensive, complex, and time-consuming maintenance and repair. This would be especially valuable in measurement or monitoring devices in environments of elevated pressure or temperature, or those which are difficult to access.

Other forms of adaptive performance can be considered from the perspective of the propagating ultrasound waves. Through adaptive signal shaping techniques such as beamforming, it is possible to tailor signal strengths in particular directions, thus optimising the performance of arrays of ultrasonic transducers, for example, in nondestructive testing and evaluation. There are also implications for artificial intelligence and its role in ultrasonic technologies. Therefore, the purpose of this book is to showcase the various dimensions of adaptive ultrasonics, through relevant transducer design and manufacture approaches, materials selection, and signal processing strategies. In terms of materials engineering, this book focuses on two broad classes, where one is shape memory alloys, with a particular focus on nickel titanium (Nitinol), and the second is active acoustic metamaterials. This book will demonstrate how these classes of advanced material can be used to deliver tailored and controllable dynamics, where some practical approaches for implementation are presented.

Through adaptive ultrasonics, it will be possible to deliver devices and systems with potentially higher performance capabilities, including adaptive dynamic features and remote or self-healing, to establish a new generation of smart and multifunctional technologies. For example, shape memory alloys and other classes of smart materials can theoretically be trained to react to changes in parameters such as temperature, magnetic or electric field, force, pH, and, in some cases, light. The field of ultrasonics is broad, but it is evidently progressing to a new phase that bridges the boundaries between ultrasonics, materials science, and computer science. We are continuing to discover more about how we can tailor the properties of advanced materials to introduce exciting new dynamic behaviours in ultrasonic devices which have thus far been difficult to achieve or impossible. On the signal processing front, there have been notable developments relating to the control and shaping of ultrasonic wave fields, with significant implications for industrial measurement. Therefore, understanding how to control the response of a system, based on the available system parameters or the functional requirements, is key for adaptive ultrasonics. This is not simply varying a property such as amplitude, but rather engineering a system to adapt to new environments or processes, either in an independent or automatic modality, or through a designed and controlled action.

Today, the field of ultrasonics is extremely varied, comprising industrial and medical innovations across orthopaedic surgery, medical diagnostics through micromanipulation, sonochemistry, therapy, geological exploration and mining, manufacturing, and nondestructive evaluation for the nuclear, energy, and aerospace industries. Generally, an ultrasonic device can typically be regarded as either a sensor, fundamentally a reactive input system, or an actuator, an active output system. Ultrasonic sensors are integral for nondestructive testing and evaluation, where they include piezoelectric and electromagnetic acoustic transducers, with longitudinal, shear, and flexural mode devices being particularly common. Ultrasound waves are generated and propagated through a medium, before they are detected, where the properties of these waves provide information about the medium or target. Ultrasonic actuators can include a variety of excitation mechanisms, where, for example, they can be piezoelectric or magnetostrictive, generating high-amplitude vibration motion of an end-effector in a tuned mode at ultrasonic frequencies. These vibrations can be used to penetrate or disrupt the form of different materials including geological, biological, and composite targets. Common ultrasonic actuator devices which are used commercially include transducers for soft or hard tissue surgeries, utilising piezoelectrically activated surgical blades, ultrasonic welders, and therapeutic transducers. There are also devices that are tailored for both sensing and actuation. For example, there are reports of high-intensity focused ultrasound (HIFU) therapy using image guidance, for example, in the work of Ning et al. (2020), optimally delivering the required high-intensity ultrasound at the target site. In these devices, though the sensor and actuator elements are distinct, they are configured into a single system. Many HIFU devices exclusively deliver therapeutic characteristics using a separate imaging process such as magnetic resonance imaging (MRI) to first determine the nature of the anatomical target. Enabling non-intrusive transformation from one to the other is a defining characteristic principle of adaptive ultrasonics.

The focus of adaptive ultrasonics in this book is divided into two key themes, where the first is materials, and the second is signal processing. The key motivation for the former in terms of this emerging research area is to understand how we can integrate different classes of material with common configurations of ultrasonic device. Many opportunities are presented by the transformational features of advanced materials such as shape memory alloys and metamaterials, but their complex properties mean that we can tailor our approach regarding how we undertake transducer design and fabrication. In terms of the latter, the aim is to showcase some common approaches to implementing adaptive signal processing, with links to artificial intelligence approaches where relevant. It is intended that the content of this book will serve as an informative foundation for future innovations in adaptive ultrasonic technologies, and as an instructional text for those researchers interested in a broad overview of some key practical aspects of ultrasonic transducer design and fabrication.

1.2 TARGET AUDIENCE

The content of this book is intended as a primer, of general interest to both academic researchers and industrial professionals and engineers and scientists. It is anticipated that the principles of adaptive ultrasonics contained here can be applied to transducer design, fabrication, and operation for a wide range of applications, and to further our knowledge on advanced materials and their behaviours. This book should be of interest to anyone engaging in the field of ultrasonics, either as a student or as a qualified practitioner. This includes, but is not limited to, those with specialisms in medical surgery, including orthopaedics; industrial manufacturing and processing methods such as ultrasonic welding; sensing applications across medicine and industry, including flow measurement; and other sensing architectures, for example, those based on metamaterials. This book aims to explore the practical design considerations which readers should find useful for the design and manufacture of the next generation of advanced ultrasonic devices and systems, presented in a logical and methodical format which is intended to be accessible for all those at graduate level and beyond. It is also intended that the lessons learned from this text can be taken further, in new directions and in the integration of a variety of new materials in ultrasonic transducers and novel concept device designs.

Fundamentals of advanced materials and the principles of ultrasound, including transducer design, should be of particular interest to graduate students and researchers in both academia and industry. The integration of certain shape memory materials and metamaterials with ultrasonic devices should be of interest to device designers, and the content provided on artificial intelligence and ultrasonics, and adaptive signal processing strategies, should be of interest to the general reader. The subject matter is interdisciplinary, forming links between dynamics, materials science, manufacturing, and computer science. This book will hopefully serve as a resource for those with interests in either and therefore broaden horizons towards new possibilities in the field.

1.3 ULTRASONIC TRANSDUCERS

Ultrasound is the term given to sound waves with frequencies above the conventionally accepted limit of human hearing, those inaudible around and above approximately 20 kHz. Ultrasound can be considered as a natural extension of the science of acoustics, where many of the foundational principles are relevant. The terms *ultrasound* and *ultrasonics* are often used interchangeably, particularly in some common descriptions and definitions of the field, but here the application of ultrasound can be regarded as ultrasonics.

The precise frequency limits which we traditionally associate with ultrasound are somewhat arbitrary. The limits associated with our capacity to hear as humans can be directly derived from the physical characteristics of our ears and our abilities to process the physical signals associated with sound. These natural limits give rise to the traditionally accepted lower band limit of 20 kHz, which can noticeably reduce with factors such as age or ill health. It should also be noted that many animals can hear far beyond this lower limit, in some cases as high as 300 kHz, where some classes of moth have been known to communicate at such frequencies (Jan, 2023). This far exceeds the most famous example from the animal kingdom, the bat, where generated calls can be in the 16–150 kHz region (Neuweiler et al., 1980). There are undoubtedly other natural phenomena, including animals, which involve very high frequency sound, from which we can learn and apply to our engineering approaches to ultrasonics. Such *biomimetic* strategies have always had an important place in ultrasonics, the growth of which is now somewhat accelerating, and will be addressed later in the book. Many of the ultrasonic applications we deal with today are in the kilohertz or megahertz ranges, and when frequencies of 1 GHz are reached and exceeded, this is commonly referred to as the hypersonic regime (Cheeke, 2010).

As a technology, ultrasonics can either be applied for lower power or amplitude applications such as sensing, where only a few volts may be required to provide the excitation signal to a transducer, or for higher power or amplitude processes including ultrasonically assisted cutting. The latter is typically classified under the general term of power ultrasonics, encompassing the spectrum of destructive processes imparting physical changes or effects to a target system. As the most common, these include surgery (bone or soft tissues), welding, joining, drilling, food cutting and processing, and any application requiring the generation of cavitation activity. A popular use of cavitation is in ultrasonic cleaning, for example, in the descaling of boat hulls. The former is more commonly associated with lower amplitude, and in many cases though not always, at higher frequencies than those utilised for power ultrasonic applications. These include sonar, imaging, flow measurement, and nondestructive testing and evaluation-based inspection of materials for phenomena such as cracks or internal structural flaws. From this point, this brief section aims to provide a concise overview of ultrasonics, where the reader is invited to read more expansively on the numerous topics that are introduced, using the expansive bibliography provided at the end of the book.

Whilst some familiarity with ultrasound and transducers is assumed, a general overview of a typical ultrasonic measurement system can be outlined. A transducer is a device which converts energy from one form into another. An actuator typically transforms an electrical input signal into mechanical output motion through the energy supplied to a suitable converter component in the system, such as a piezoelectric material. A sensor can convert mechanical vibrations, or electrical signals, into a response as desired, either a mechanical output or another electrical signal. A transducer is generally driven through a signal generator where specific drive

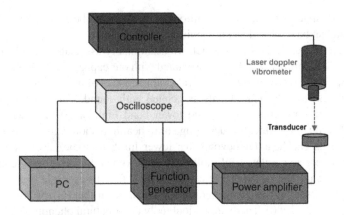

FIGURE 1.1 A typical experimental ultrasonic measurement system, consisting of a transducer connected to a generator. A laser Doppler vibrometer is included here to show how optical measurement and monitoring can be integrated with the system.

conditions can be configured, including voltage, cycle type (e.g., sinusoidal, chirp, or burst), number of cycles in a pulse if relevant, and the spacing between pulses in the excitation. It is often necessary to include a power or conditioning amplifier in the system, depending on the amplitude requirements and characteristic electrical conditions including impedance. A typical configuration of ultrasonic measurement system is depicted in Figure 1.1, which includes an optical measurement instrument called a laser Doppler vibrometer, useful for the analysis of physical vibration amplitudes of structures. The schematic illustrates the key measurement components of interest that are common across many applications.

Here, the common types of ultrasonic transducers can be introduced. There are many types of ultrasonic devices available, with a variety of custom or tailored configurations depending on the application reported in the literature. It is therefore difficult to classify them purely according to their construction or appearance. Depending on the configuration and target performance of interest, an ultrasonic transducer can incorporate a piezoelectric material (where an electric charge across the material delivers a mechanical stress, an effect which is generally commutative), a magnetostrictive material (conceptually similar to the piezoelectric effect but in this case relies on magnetism), or be excited through various mechanisms, such as electromagnetically via a coil (electromagnetic acoustic transducers, or EMATs), or via a tailored electrode (in the case of a capacitive micromachined ultrasonic transducer). These materials can all be built into an ultrasonic transducer in different ways, and these determine the operating characteristics of the device.

For example, piezoelectric materials can be fabricated into different shapes and sizes, and depending on their geometry and poling condition, they will vibrate in a particular way in response to an applied electric charge. These piezoelectric coefficients, some of which will be detailed later in the book, denote the change in volume of the material under the influence of an electric field applied to it. These coefficients are associated with different dimensions, giving a transducer designer guidance on how a device can be optimised for its dynamic performance. In the case of the Langevin transducer, this configuration is typically assembled using rings of piezoelectric material. The piezoelectric coefficients associated with their thickness modes will in part determine the geometry of the transducer to be tuned to a particular frequency and mode shape.

Common configurations of ultrasonic transducer, both for lower and higher amplitude applications, are illustrated in Figure 1.2. Through the inspection of these configurations in more detail, with a selection of others for a broader context

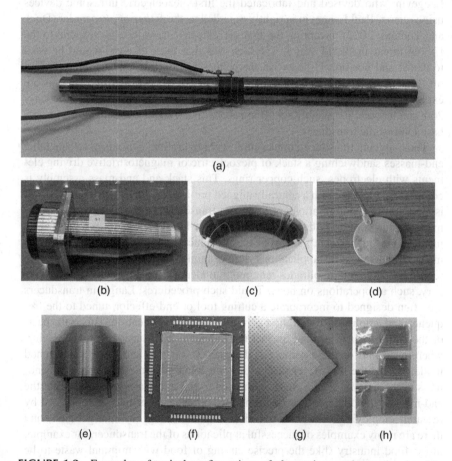

FIGURE 1.2 Examples of typical configurations of ultrasonic transducer, showing (a) the Langevin (Liu et al., 2024a), (b) the Tonpilz (Pyo et al., 2021), (c) the Class IV flextensional (Zhou et al., 2018), (d) the Class V flextensional cymbal (Shim and Roh, 2019), (e) the flexural (Feeney et al., 2018a), (f) the piezoelectric micromachined ultrasonic transducer (Joshi et al., 2023), (g) (an array of) Capacitive micromachined ultrasonic transducer micromachined ultrasonic transducers (Manwar and Chowdhury, 2016), and (h) electromagnetic acoustic transducer coils (Jiang et al., 2023). All images are provided here to give the reader a qualitative and general idea of the structural configuration of each class of transducer. It should be noted that variations on each of these can be found in the literature, and that all images here have been reprinted and/or adapted under the CC-BY 4.0 licence.

relating to what designers may be targeting, a concise overview of each transducer type is provided below. There are many configurations of transducer, but many of the popular applications of ultrasound will likely incorporate one or more of these types.

1.3.1   THE LANGEVIN TRANSDUCER

The Langevin transducer is named after the eminent French physicist Paul Langevin, who devised and fabricated the first piezoelectric ultrasonic devices using quartz. Paul Langevin is rightly regarded as the *father of ultrasonics* (Duck and Thomas, 2022), inventing the first piezoelectric transducer to respond to the U-boat threat in World War I. The idea was that these devices would be used to detect and position U-boats via the echoes from ultrasound waves reflecting off the targets. The early developments in ultrasound and their applications are explored in greater detail in Chapter 2, but this field began in earnest in the early 20th century, and it arguably set off at pace with the advent of Langevin's quartz-based ultrasonic transducer.

The Langevin transducer, in its simplest configuration, consists of two metallic end-masses sandwiching a stack of piezoelectric or magnetostrictive driving elements with electrodes, such copper rings. This stack and end-mass assembly is compressed together with a centrally aligned pre-loading (or pre-stressing) bolt. It is common for configurations of Langevin transducer to be designed to accommodate a range of amplification components, or horns. These horns significantly increase the quantifiable gain from the transducer, optimising it for a range of medical and industrial applications. For example, such horns are necessary to realise the vibration amplitudes required to undertake ultrasonically assisted surgery, such as operations on bone. To aid such procedures, Langevin transducers are often designed to incorporate a cutting tool or end-effector, tuned to the frequency and amplitude of vibration as required by the application. This is the basis of the entire transducer concept that the Langevin transducer is a tuned device, where each part of the assembly is designed to operate and vibrate to a tuned mode. Within the transducer design, it is customary to integrate a heavier end-mass at the rear, or a back mass, compared to the material used to fabricate the end-mass at the other end, or the front mass. Power ultrasonics is dominated by research employing ultrasonic transducers in the Langevin configuration, and there are many examples of successful applications of the transducer, for example, in the food industry (like the precise cutting of food with minimal waste to be packaged) and in manufacturing (such as ultrasonic wire bonding), beyond the applications of ultrasonically assisted surgery and underwater sonar.

The Langevin transducer presents the ideal opportunity for realising a new generation of adaptive ultrasonic devices, but it also brings about significant challenges. It is a configuration relying on tuned components, and it can be a difficult transducer to design and fabricate, in terms of both labour time and expense. It generally requires specialist knowledge to tune and operate, with much derived from experience. Nevertheless, its success as the core transducer of choice across medical and industrial power ultrasonics for over 100 years demonstrates its versatility and effectiveness in delivering ultrasound for the benefit of society in a multitude of areas.

1.3.2 THE TONPILZ TRANSDUCER

The term *Tonpilz* is derived from a combination of the German words for *tone* and *mushroom*, where the former term relates to its use in sound applications, and the latter term for its geometrical shape. It can be regarded as an optimised form of the Langevin transducer because it consists of a sandwich structure of a piezoelectric or magnetostrictive driving element in between two end-masses. These end-masses comprise a comparatively light front mass which acts as the radiator, with a relatively heavy back, or tail, mass. It is regarded as a relatively straightforward and inexpensive transducer to design and manufacture (Rouffaud et al., 2015), and common choices of material for the head and tail masses include aluminium and brass, respectively. Recent research has investigated the incorporation of lead-free piezoelectric materials in the transducer configuration (Rouffaud et al., 2015), as a progression from the use of conventional compositions of lead-based piezoelectric materials. Further reference to piezoelectric ceramic material choice will be made later in this book.

The Tonpilz transducer tends to be utilised as an electroacoustic transducer for underwater applications such as sonar. It can be either a projector of ultrasound or a receiver, where in this case it would be more accurately termed as a hydrophone. It is usually embedded in arrays of several transducers and can be fabricated to operate at a range of frequencies. One key benefit of incorporating this transducer in array configurations is that it makes possible beam-steering or beamforming, through the control of the phase response of the transducers relative to one another.

1.3.3 THE FLEXTENSIONAL TRANSDUCER

In classical terms, the flextensional transducer is a distinct class of electroacoustic transducer, traditionally for underwater applications and originally referred to by Toulis in early research into novel concepts for sonar-type devices (Rolt, 1990). The primary mechanism involved in how a flextensional transducer operates is through a combination of flexural and extensional motion, realised via a thin metallic cap or shell. The transducer is typically composed of one or more metallic shells, depending on the class of flextensional transducer, where the vibration motion of these shells is stimulated by the application of an electrical voltage to a piezoelectric stack. This piezoelectric stack is embedded at a key location in the device structure, usually at an optimal position that generates the highest vibration amplitudes for a given excitation condition.

In general, flextensional transducers align to one of seven major classes which can be viewed in the work of Sherman and Butler (2007), differentiated based on their shell geometry or configuration and intended end-use application. It should be noted that in this book, the terms 'shell' and 'cap' may be used interchangeably.

Irrespective of the transducer class, a flextensional transducer, as a system, will consist of the following principal components: the shell(s); the active (usually piezoelectric) stack; and the medium in which the transducer is operating. Some of the early developments in the flextensional transducer, beginning with key innovators such as Hayes and Toulis in the early to mid-20th century, are detailed in Chapter 2 and are included to provide a comprehensive background regarding from where contemporary device designs originated.

The configurations of flextensional transducer of most relevance and interest to the contents of this book are the Class IV and an adaption of the Class V, known as the cymbal transducer. These types of flextensional transducer can both be seen in Figure 1.2. The principal reason is that they have received significant attention in recent years at the interface of sonar and power ultrasonic transducers for alternative applications, including ultrasonic cutting. Flextensional transducers are traditionally designed for sonar and underwater sound propagation, but there have been significant efforts in recent years to transition these concepts towards other areas including surgical cutting and energy harvesting. Very recently, new design concepts of flextensional transducers with adaptive properties have been introduced, such as through embedding shape memory materials into their structure to enable frequency control. This will be expanded upon later in the book, but the principal point is that the Classes IV and V are ideal candidates for adaptive ultrasonic transducers given their relatively straightforward structural composition and few components required for their assemblies.

1.3.4 THE FLEXURAL ULTRASONIC TRANSDUCER

The flexural ultrasonic transducer (often known as the FUT) sits in that novel and uncommon space between being an exceptionally widely used configuration of ultrasonic transducer and also one which until the mid-2010s was remarkably under-studied, under-utilised, and poorly understood. The flexural ultrasonic transducer has long been the transducer configuration of choice for the automotive industry, in their car parking proximity sensing systems. The transducer can operate as either a transmitter (or projector) or a receiver (or detector) of low-frequency ultrasound and is an exceptionally efficient device, where it can propagate ultrasound over several metres and beyond for only a few volts of excitation voltage. It is also relatively inexpensive and straightforward to fabricate, since commercially available configurations can be purchased for a few GBP.

However, there have been limitations to this transducer configuration. Every commercially available transducer adheres to a broadly similar fundamental design concept. A piezoelectric ceramic disc is bonded with an epoxy resin to the underside of a metallic plate (usually aluminium) which is generally in the order of 10–20 mm in diameter. This structure is enclosed in a metallic housing, again which tends to be aluminium, or the same material as the plate, with silicone-type material used to produce a nominally air-tight backing. As a voltage excitation is applied to the piezoelectric ceramic, its resulting vibrations stimulate plate modes in the aluminium or metallic plate, thereby producing the ultrasound. The frequency and other dynamic characteristics are dependent on, and strictly dominated by, the material choice of the plate and its geometrical dimensions. Most commercially available flexural ultrasonic transducers operate around 30–40 kHz in their fundamental mode of vibration.

The principal advantage of the flexural ultrasonic transducer, unlike some other configurations of ultrasonic transducer such as the piezoelectric micromachined ultrasonic transducer (PMUT), is that it does not require any matching layers to be integrated with the vibrating plate (sometimes referred to as a membrane, though with its flexural rigidity it can be designated as a plate). This is a significant advantage because it reduces both the cost and complexity of the fabrication process. It is also one of the reasons it is popular in automotive applications, because their relatively straightforward design and fabrication makes them highly conducive to mass production. Of course, there are drawbacks, one of which is the achievable structural consistency from one flexural ultrasonic transducer to another, which limits dynamic performance. Nevertheless, the flexural ultrasonic transducer operates a little like the human eardrum in the way it detects ultrasound, and recent research is uncovering new and exciting applications for this class of transducer in the future.

At present, commercially available flexural ultrasonic transducers are all designed for operation at ambient atmospheric pressure and room temperature conditions. Significant early research has been undertaken to understand this and suggest developments on the transducer design, including Dixon et al. (2017), Feeney et al. (2018a), and Feeney et al. (2019a). There has also been research to increase its operational or resonance frequencies, which leads on to another advantage of the flexural ultrasonic transducer. It operates via either axisymmetric or asymmetric plate modes, but depending on the drive conditions, any of these can theoretically be selected (Feeney et al., 2018a, 2018b). The challenge is optimising the transducer design to efficiently operate at higher order modes of vibration, if so required. The opportunity for the flexural ultrasonic transducer to be a true multi-mode device has yet to be fully realised, but its potential has already been demonstrated.

The flexural ultrasonic transducer has been trialled in prospective new applications including flow measurement (and due to its assembly not requiring matching layers, its nature is that it would be able to measure in both liquid or gas), anemometry, and proximity measurement in pressurised environments. Investigations that are commercially aligned, and impact case studies in general, need to be progressed, but research outcomes thus far in all these areas have indicated the real potential for true commercial potential. There have also been developments in flexural ultrasonic transducer design for environments of elevated temperature, postulated to be revolutionary in the future for industrial processes including flare gas measurement.

Another limitation of the flexural ultrasonic transducer in its classical form is that it possesses a relatively long resonant ring-down time of its vibration response because it is effectively a narrowband device. Some recent research has mitigated this, for example, in the engineering of novel active damping methods (Dixon et al., 2021). These have demonstrated how the flexural ultrasonic transducer response can be controlled, indicating that resonant ring-down can be mitigated, thus making the transducer suited to environments of lower volume and conditions of shorter response times, further expanding on the possible applications in the future. Current research is also focusing on novel configurations of plate and alternative materials for these to further optimise the response of the transducer. The reader is directed to Appendix A1 for a full analytical equivalent model of the dynamic response of the flexural ultrasonic transducer, but which can generally be considered for a variety of ultrasonic transducer configurations.

1.3.5 THE PIEZOELECTRIC MICROMACHINED ULTRASONIC TRANSDUCER (PMUT)

The piezoelectric micromachined ultrasonic transducer, or PMUT, is a popular configuration of transducer in contemporary ultrasound applications. It is a microelectromechanical system (MEMS)–type device where the principal governing mechanism of operation is through a flexural vibration mode of a membrane, to which a piezoelectric-type material is attached, generally as a thin film. An advantage of the PMUT concept is that it avoids the need for using bulk piezoelectric ceramic materials, such as discs, which often exploit the thickness modes of these ceramics, often with lower bandwidth than what is achievable using a PMUT. They can also require low voltages to operate efficiently, can be manufactured very small compared to transducers incorporating bulk ceramics, and with the right type of matching layers, can be tailored for operation in liquids and gases, such as air and water (Roy et al., 2023).

As reported by Roy et al. in 2023, the PMUT can be traced back to the late 1970s, from the creation of sputtered zinc oxide (ZnO) layers for sensor systems by Shiosaki et al. (Roy et al., 2023). Through the 1980s and into the 1990s, different teams of researchers demonstrated how thin film technology could be used in the construction of PMUT devices, using a range of materials including ZnO and piezoelectric sol–gel. Further to these developments, PMUTs began to be exploited for applications including imaging in water, such as the work of Bernstein et al. in 1997. Into the 2000s, it was possible to deposit even thinner layers, and other materials came to prominence including polyvinylidene fluoride (PVDF) and aluminium nitride (AlN). Intercardiac tomography was even achieved as recently as 2014 using a piezoelectric lead zirconate titanate (PZT) PMUT. Development has further progressed recently to more complex active material layers and more intricate and smaller-scale applications, where PMUTs are often integrated as arrays. The flexibility of the materials is a topic of current interest, as has the development of piezoelectric and triboelectric nanogenerators (PENGs and TENGs) in the 2010s, which are effectively energy harvesting devices.

The basic construction of a PMUT consists of two major parts, the device structure incorporating the membrane, and the active piezoelectric part, which, for example, can be a thin film of piezoelectric material directly deposited on to the membrane (Roy et al., 2023). There are two distinctions of PMUTs, depending on the scale of the device being fabricated. One type is regarded as membrane based, and the other is plate based. Plate PMUTs are generally thicker, where a DC voltage applied to the piezoelectric material generates a strain which then leads to a bending of the constrained plate. The membrane-based PMUT comprises a membrane which is effectively pre-stressed, where an AC voltage applied can be used to instigate the membrane vibration. There are many configurations of PMUT described in the scientific literature, but popular active material layers of interest include those based on ZnO, AlN, and PVDF.

PMUTs can operate either as transmitters, receivers, or as transceivers, similar to the operation of a flexural ultrasonic transducer detailed in the preceding part of this section, and they can be tailored to operate in different vibration modes or at a range of frequencies as required by the application, commonly into the MHz range. A key advantage of the PMUT is that since it is a MEMS-type device, it facilitates the manufacture of ultrasonic transducers at small scales. Leading on from this, ultrasonic transducers which are smaller in scale than what we have been able to produce up to this point, realised from several developments including advanced manufacturing technologies, including the additive manufacturing of many different types of material, enable multifunctionality, lower levels of power consumption, lower noise, and the capacity for batch manufacture (Jung et al., 2017).

1.3.6 The Capacitive Micromachined Ultrasonic Transducer (CMUT)

The capacitive micromachined ultrasonic transducer, more commonly referred to as the CMUT, is the other major class of MEMS-type ultrasonic transducer which is popular in contemporary sensor systems. The CMUT was conceived in the 1990s (Khuri-Yakub and Oralkan, 2011), being a much more modern innovation compared to its (arguably very different) piezoelectric counterpart. The fundamental basis of operation of a CMUT is parallel plate capacitance, where a thin plate is supported over a vacuum space, thereby producing an electrode surface. Another electrode is created by the inclusion of a substrate, and if a DC voltage is applied across these two electrode surfaces, then the electrostatic force which is generated attracts one electrode towards the other. There is also a mechanical restoring force due to the mechanical properties of the plate, and so by the application of an alternating voltage, ultrasonic waves can be produced.

Since the CMUT can act as a receiver, any ultrasound wave impacting this membrane generates a flexing motion which directly influences the capacitance, thus providing the detection mechanism (Sautto et al., 2017). As a transmitter, an alternating voltage is applied to the plate and membrane configuration, where an electrostatic force is produced that subsequently generates the vibrations of the CMUT's membrane, aligned with the physical characteristics of that membrane. CMUTs typically exhibit relatively broad bandwidth and their fabrication can be relatively straightforward. However, relatively large biasing voltages can be required to drive a CMUT, and this can be a notable disadvantage.

CMUTs can be designed to operate far into the MHz range, notably in the works of Martinussen et al. (2009) and Gross et al. (2015), and they can be engineered to exhibit broad bandwidth. Like the PMUT, they are MEMS-type devices and so can be readily miniaturised to a scale suitable for many advanced electronics applications and complex system architectures. They are commonly used for imaging applications, in microscopy, for proximity measurement given they can couple well to air, and in ultrasonic flow measurement.

1.3.7   THE ELECTROMAGNETIC ACOUSTIC TRANSDUCER (EMAT)

A class of transducer which has been particularly successful in the non-contact, nondestructive testing and evaluation field is the electromagnetic acoustic transducer, more commonly referred to as the EMAT. According to Hirao and Ogi, authors of a key text in the field of EMATs (Hirao and Ogi, 2013), the historical origins of this configuration of transducer are a little unclear, but there are key moments recorded in the scientific literature which allude to some pathway towards where the discipline is in modern times. The progress made in detection using magnetic elements in the 1930s by Wegel and Walther certainly laid the groundwork for later developments in the use of electromagnetic waves to generate and detect shear and longitudinal modes (Wegel and Walther, 1935; Hirao & Ogi, 2017; Xie et al., 2020; Martinho et al., 2022).

The basic principle of an EMAT is that it is a coil of wire which is used to generate a dynamic electromagnetic field in the surface of a material, which evidently must be conductive. EMATs are capable of both transmitting and receiving ultrasound and are inherently useful as non-contact devices, particularly for those applications where not being in direct contact with a target structure is of the utmost importance. This can be due to a target structure being excessively hot, difficult to access, or in motion (Petcher et al., 2014). The ability of EMATs to generate and detect ultrasound relating to very hot structures should be noted as a key advantage, because temperature can be a key restrictor in the application of piezoelectric-based transducers where degradations in dynamic and mechanical performance can develop.

The design challenges associated with EMATs are significant because they require high levels of precision in how the wires in the transducer are coiled. EMATs generally require a permanent magnet which creates the static magnetic field needed for the device to operate. Fundamentally, the EMAT works via the inclusion of a metallic coil as part of the transducer structure. The static magnetic field is produced by magnets, and this field interacts with a field which comes from the electrical coils, at a higher frequency. A Lorentz force is then generated which directly influences the lattices of a target material, thus producing ultrasound via the mechanical waves induced. The inverse behaviour also occurs, where these waves that come into contact with the generated magnetic fields can lead to the generation of currents in the coils of the EMAT.

Although this transducer configuration is technically non-contact, it cannot be used too far away from the target structure, or the ultrasound cannot be generated given the reliance of the electromagnetics on the static magnetic field and the presence of the coil, which itself is often fabricated from a material such as copper. A parameter known as 'lift off' is therefore important to consider in how an EMAT is used. Advantages of the EMAT, noting the cited literature in this section including Hirao and Ogi (2013) and Petcher et al. (2014), include the fact that no couplant is required, given the lift off capability, and target surfaces therefore do not generally need to be prepared or polished to a particular standard for high-quality measurement. It should be noted though that EMATs can exhibit lower transduction magnitudes than alternative configurations such as piezoelectric-based transducers.

1.3.8 CONTACT TRANSDUCERS

This category is defined to include single-element transducers, such as wedge transducers, which incorporate materials like piezoelectric ceramic discs. These devices are often used in nondestructive testing and evaluation, or for applications such as clamp-on flow measurement. They typically utilise a longitudinal or shear wave mode in direct contact with a target, depending on the transducer construction and the intended application. The main operational consideration for a contact transducer is that it is intended for use in direct contact with a target material, most commonly a structure for inspection in a nondestructive testing and evaluation application. One example would be the inspection of rails in the railway industry.

Contact transducers are often used by hand, directly on the target surface of interest by a user. However, advances in robotically enhanced ultrasound inspection are taking contact ultrasound measurement beyond the human user and into newer application spaces (Vasilev et al., 2021), including those involving environments of elevated temperature and pressure, or those which are hostile. Given there is direct contact between the transducer and the subject structure, measurement and inspection processes can often require the use of couplant materials to ensure there are minimal quantities or sizes of air gaps between the transducer face and the target structure. This is to prevent unwanted reflections or energy loss which would cause a reduction in the quality of the measurement. Common couplant agents include silicone grease, some oils, and occasionally liquids including water.

1.3.8.1 Immersion Transducers

Ultrasonic transducers which are specifically designed for operation in a fluid of interest, in this case water, can be referred to as an immersion transducer. There is nothing particularly unique regarding the materials or construction of an immersion transducer, but they are notable for their specific use in the industrial measurement community where monitoring and measurement of structures which are submerged in water is of importance. In water, an immersion transducer propagates ultrasonic waves in a longitudinal mode, emanating from a single element which is matched, in terms of its acoustic properties, to water. These transducers are commonly piezoelectric based and often incorporate a level of scanning or focusing to enable a rapid and accurate picture of a test subject to be generated.

Immersion transducers will often consist of an active material, like a piezoelectric ceramic, and include matching layers which are acoustically close in terms of impedance to that of water and far lower in terms of acoustic impedance than those of bulk piezoelectric ceramic materials. This allows the levels of sensitivity in operation needed to achieve reliable and high-performance ultrasound measurement, utilising both scanning and focusing capabilities where relevant. One important factor to note is that these matching layers can commonly be found to be specified as *quarter wavelength,* allowing optimal ultrasound transmission (Suñol et al., 2019). The acoustic impedance for the chosen matching layer is ideally between that of the piezoelectric active material and that of the surrounding fluid. It is customary to introduce additional layers as required, to enhance the performance of the transducer (Suñol et al., 2019). Piezoelectric transducers in this case are often fabricated with a resonance frequency which is set by the thickness of the piezoelectric element, that for this classification is commonly a disc.

In terms of design principles, it is important to ensure that the piezoelectric material and its matching layers are configured to facilitate an optimal level of energy transfer with water, given the environment in which immersion transducers are commonly utilised. It is also customary to house the device using corrosion and erosion resistant, or resilient, materials. These can include materials such as alloys based on stainless steels.

1.3.9 PHASED ARRAY TRANSDUCERS

Phased arrays are effectively a series of transducer elements configured to operate together or in phase, in a singular system. They are distinguished from other, single element, transducers because they are often used in a similar manner to deliver ultrasound to a target. These arrays can be constructed from individual elements of the types already outlined in this chapter, including flexural type or PMUTs. The principal advantages of using phased array configurations are that, providing they are fabricated with the required precision, they can deliver optimal amounts of ultrasonic energy to a target; they can be used to steer ultrasound beams; for example, in the case of mid-air haptics (Hasegawa and Shinoda, 2018; Price and Long, 2018), they can be used to scan or obtain data over a wide surface area and potentially rapidly; for example, in flow measurement (Kang et al., 2017, 2019); or be used to focus ultrasound in a particular way or in specific pulse shapes, depending on how the individual array elements are globally phased and configured.

Phased arrays are most popular in nondestructive testing and evaluation, and whilst they offer the advantage of being able to inspect rapidly and over wide areas, advances in robotic inspection are providing alternative capabilities to some array designs (Macleod et al., 2016; Khan et al., 2019). There are complexities involved in the design and fabrication of phased array transducers. Arrays require similarity in terms of dynamic performance, key being resonance frequency, of each individual element, and little crosstalk between these elements (therefore, the dynamics of one element influencing those in proximity). Overcoming crosstalk can be achieved by the design of a rigid baffle into which transducer elements are housed (Kang et al., 2017, 2019), which is a significant mechanical engineering challenge. Furthermore, ensuring that the driving network for the phased array is configured correctly is another major challenge, and often the controllers needed for such systems are complex and expensive. Depending on the application, phased array transducers may incorporate a significant number of elements, all of which need to be wired up and operated in phase for the system to work effectively. Phased arrays are typically known to accommodate anywhere from a few elements into the hundreds, depending on the operational performance required.

Nevertheless, a robust and high-performance phased array system can provide remarkably high-quality ultrasound measurement data, and it can be tailored to inspect a wide range of measurement subjects, including metallic parts where inspection at different depths is required. A phased array system could be used to focus ultrasound at different depths, depending on the phasing and configuration of the individual transducer elements in the array. Importantly, there is now versatility in the configuration of elements in a phased array system. For example, some can be commercially acquired in convex shapes, which are particularly good for depth-based inspection or in haptics technologies (Morisaki et al., 2023), or they can be in circular or annular patterns which are beneficial for inspecting structures which are not flat or have some irregularity in their dimensioning, or more commonly flat or square arrangements.

1.4 TRANSDUCERS AND WAVE PROPAGATION

In general, ultrasound as a field can be considered as two broad areas, with one low-intensity and high-frequency, and the other high-intensity and low-frequency (Gallego-Juárez et al., 2023a). The cut-offs for these categories are not always straightforward to define, but in general low-frequency ultrasound is commonly associated with those in the 20 kHz–100 kHz frequency range. Most power ultrasonic processes are configured for this frequency range, but there are some exceptions to the rule. For example, low-intensity ultrasound generated from flexural-type sensors, such as those which would be commonly found in conventional automotive parking systems, can typically operate around 40 kHz (Feeney et al., 2017a, Feeney et al., 2017b). Thus, it is also not unusual to find low-intensity and low-frequency systems, and vice-versa.

In some configurations of ultrasonic transducer, the output amplitude, such as the displacement, must be optimised because there is not a sufficient level of gain. The bolted Langevin transducer, arguably the most popular configuration of ultrasonic device as detailed in this chapter, typically incorporates a horn on the end which is tuned to the transducer's resonant mode of interest. This is often the first longitudinal mode of vibration, though this is not always the case and depends on the nature of the motion of interest. Fundamentally, the purpose of attaching a tuned horn to a Langevin transducer is because without this, the device would not be able to deliver the amplitudes required at low ultrasonic frequencies necessary for power ultrasonic applications.

With the transducer and the signal generator, often with some form of matching network to optimise the energy transfer into the ultrasonic transducer and to compensate for mismatches in electrical impedance, the ultrasonic transducer can then be operated with drive parameters suited to the target application of interest. A general introduction to ultrasonics and key transducer classes have been presented thus far, and although some of the governing principles of ultrasonic wave propagation are well established, it is necessary to provide a concise outline here.

The first point to note, irrespective of whether an ultrasonic transducer is designed to propagate ultrasound waves towards a target through a gas or liquid fluid medium (such as in flow measurement or proximity sensing), a solid-like medium (e.g., in nondestructive testing and evaluation of metallic structures), or if the cyclic motions of a vibrating structure at ultrasonic frequencies are to be exploited for destructive purposes, such as cutting or welding, ultrasonics always involves the generation and propagation of ultrasound waves. Ultrasound waves are just sound waves but at frequencies above the range of human hearing. Sound can be regarded as a series of vibrations or pressure changes in a medium which then manifest as acoustic, or sound, waves. These waves are mechanical in nature, where energy is transmitted through the molecular structures of the host media. In fluids, like liquids and gases, sound travels longitudinally; however, there are both longitudinal and transverse waves possible in solids, and the major reason for this is the shear modulus property of solid materials. Longitudinal waves typically manifest as localised cycles of compression and expansion, more commonly referred to as rarefaction, with shear motion a characteristic of transverse wave propagation in solids.

A key feature of sound waves is the superposition principle, where sounds can be regarded as a sum of a certain combination of sinusoidal waves, for example, if they exist at one location at a given time. This can be readily demonstrated through Fourier analysis. This superposition affects parameters including amplitude and frequency, and it is an important consideration for understanding wave propagation. The propagation of sound waves itself is governed by the acoustic wave equation, which in a general form for one dimension can be expressed as Equation (1.1), adapted from Feynman (1963).

$$\frac{\partial^2 P}{\partial x^2} - \frac{1}{c^2}\frac{\partial^2 P}{\partial t^2} = 0 \qquad (1.1)$$

Here, the P parameter is the acoustic pressure, the c parameter is the sound speed in the medium of interest, and the x parameter is the position. The acoustic wave equation is a linear expression and as shown in the equation above, it describes the relationship between the pressure and the velocity of a particle associated with the wave propagation condition, in terms of the position of that particle with respect to time. It is strictly a combination of three different relationships, those being the *conservation of mass*, the *conservation of momentum*, and the *state equation* governing thermodynamic principles.

The wave equation shown by Equation (1.1) is one-dimensional, but it can be used for several purposes. For example, in the contexts of this book, it can be applied in the modelling and simulation of vibrating rods in terms of their longitudinal and torsional modes. This is of direct relevance to modelling of a Langevin transducer if its structure is simplified to that of a series of vibrating rods. This can be readily achieved by consideration of the fact that the velocity of sound through a rod can be expressed as the square root of the ratio of Young's modulus of the rod material to its density, or specific modulus, as shown in (Field, 1931). The acoustic wave equation can also be expressed in three dimensions, though it is somewhat out of scope for this book. The acoustic wave equation in three dimensions can be used to model and analyse a range of systems, including in the study of seismic phenomena, or in nondestructive testing and evaluation applications, including more complicated structures. The mechanism by which sound waves travel through different media is illustrated in Figure 1.3, showing a generic

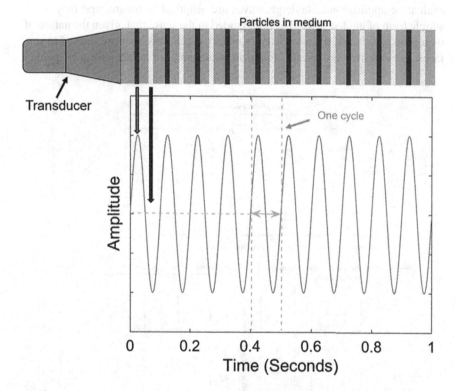

FIGURE 1.3 Propagation of a sound wave, showing (top) a transducer or sound source which generates sound through a medium, which could be a solid, gas, or a liquid. This medium contains particles which vibrate depending on factors such as the frequency and amplitude of the sound source, creating zones of compression and rarefaction. The sinusoidal wave shown is 10 Hz in the window given the designated time in seconds, where a single cycle is highlighted. The amplitude is arbitrary.

transducer which can be used to create sound waves, and a medium containing particles that mechanically vibrate in response to the energy from the sound waves travelling through. This illustration also shows an example of a sinusoidal sound wave at 10 Hz, demonstrating how alternating zones of compression and rarefaction in the sound wave propagation creates the characteristic wave, also indicating a single cycle of vibration within this. Amplitude is commonly taken between the horizontal dotted line and the wave peak.

There are several types of propagating ultrasound wave, and a selection is introduced here for context. For reference, qualitative illustrations of these waves are shown in Figure 1.4 (Alobaidi et al., 2015). The first is *longitudinal*, found in liquids and gases and where the motion of wave propagation correlates with the motional displacement and velocity of the particles within the host medium. The second major type is *transverse*, occurring perpendicularly to the direction of wave propagation and found in solid materials. *Rayleigh* waves are surface acoustic waves (SAWs) which are commonly exploited for nondestructive testing and evaluation applications. Rayleigh waves are elliptical in nature, and they are a simple form of guided wave. They are guided in the sense that, given the nature of their propagation and the associated boundary conditions, they are limited to surfaces. *Lamb* waves are strictly found in plates and are another form of guided

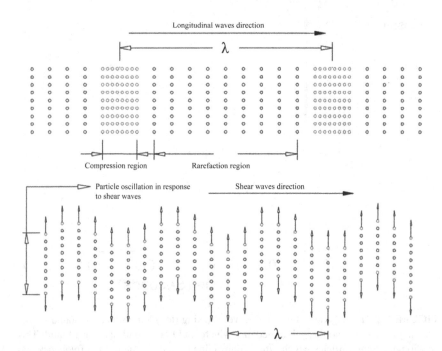

FIGURE 1.4 Wave types, showing longitudinal (top) and shear (bottom), in terms of alternating cycles of compression and rarefaction, and the wavelength λ, reprinted from the work of Alobaidi et al. (2015), under the CC-BY 4.0 licence.

wave, and they are again used for nondestructive testing and evaluation. With Lamb waves, the propagation of the ultrasound spans both the direction of the travelling wave and in the perpendicular direction.

In terms of ultrasound wave propagation in general, in a theoretical sense which is convenient for modelling and approximations for practical application, it is often customary to consider propagating waves in terms of the source characteristics. The first to raise is the *plane* wave, an idealised concept designating a series of points on a wavefront, each radiating sound omnidirectionally, which, when combined, can be equated to the wave propagation of that wavefront. This is the essence of Huygens' Principle, and it is a convenient mathematical approach to the modelling of such waves. Point sources themselves are omnidirectional in nature, often referred to as spherical waves, and are commonly used to represent underwater ultrasound sources over significantly large areas. Even though the representation is approximate, given the dimensions involved, it is usually a sufficiently accurate equivalence. Finally, if we regard sound waves in general, care must be taken to distinguish between the near and far fields of a propagating wave. Once the source of a wave is considered sufficiently far away, its behaviour can be approximated as a plane wave. However, at a distance much closer to the source, the physical effects of being close to the source of sound generation can influence and distort the wave propagation patterns, making measurements more complicated.

In discussions regarding acoustic wave propagation, there must be a reference to the principle of dispersion. Whilst this may not have direct relevance to some of the material which follows later in this book, it is a critical factor relating to ultrasound wave propagation. Dispersion is a feature of a system which describes how the velocity of a propagating wave is dependent on the frequency of the wave. In practical terms, this would manifest as an ultrasound wave splitting into separate, or component, frequencies due to the passing of that wave through a specific medium. This directly impacts the quality and characteristics of the propagating ultrasound waves, and it also leads to the difference in wave velocities which can manifest. It should also be noted that materials can also absorb sound, and this is known to be frequency dependent.

If we return to the superposition principle raised earlier in this chapter, it is known that any sound wave can be broken down into a series of separate sinusoidal waves, that when summed, produce that sound wave. In some cases, when waves with different frequencies are summed, a beating phenomenon is produced, with a distinct envelope. This beating is based on the constructive and destructive interference arising from the summation of the constituent waves, and manifests in a physical sense as a beat involving periodic changes in volume. In addition to interference, beating is also due to velocity being a frequency-dependent property, as we know given their relationships with wavelength. Therefore, a wave can propagate at its own *phase* velocity, whereas the envelope can propagate at a *group* velocity. In real practical applications, it is common to encounter dispersion due to physical geometry and material properties.

The next major consequence of the wave propagation of ultrasound in real applications is attenuation, which is the phenomenon whereby the intensity or the amplitude of the ultrasound waves is reduced, as the propagation moves through a target medium. This is very common in imaging, for example, in medical applications. Attenuation can take the form of absorption or scattering in a target medium, where the physical properties of the target material limit the reach of the ultrasound waves and their effectiveness in travelling into a target medium by a transmitter or source, and back again to be collected by a detector. It is known that higher frequency ultrasound is more commonly affected by attenuation than lower frequency, and it is therefore more of a significant problem in higher frequency applications such as medical ultrasound imaging. There are models which are often applied to understand the propagation of ultrasound waves in tissues, and these include the Maxwell and Voight models (Chen et al., 2012), both on which the reader is encouraged to read further. Through the application of the acoustic wave equation and the use of an appropriate viscoelastic model such as the Voight model for human tissue, the attenuation of ultrasound can be quantified and understood.

In general, it should be apparent that the efficacy and characteristics of ultrasound wave propagation are significantly influenced by the properties of the medium in which the ultrasonic wave is travelling. An important parameter which requires some attention here is acoustic impedance, which describes how resistant a material is to the propagation of ultrasound waves travelling through it. This is a critical parameter in the design of ultrasonic transducers because they are complex composite structures, where each different constituent material possesses its own acoustic impedance. The principle of acoustic impedance in the operation of an ultrasonic transducer is depicted in the general schematic shown in Figure 1.5.

Here, the transducer can be considered as incorporating two mediums of different acoustic impedance in its assembly, denoted as Medium 1 and Medium 2. In Figure 1.5, the solid arrow represents the sound waves from the transducer propagating through Medium 1, and the large-hatched arrow represents the acoustic energy in the system reflected to the sound source (the transducer), where the interface between the two media has prevented some of the waves from travelling. The small-hatched arrow in Medium 2 represents the remainder of the sound waves that have been permitted to pass. The angled-solid arrow indicates sound

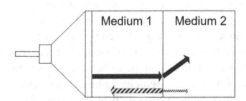

FIGURE 1.5 Simplified representation of the influence of acoustic impedance on wave propagation, showing a propagating ultrasound wave in Medium 1 (solid), some of which is then reflected (large-hatched), with the remainder propagating into Medium 2 (small-hatched). Occasionally, the direction of sound waves can change at an interface with an acoustic impedance mismatch (angled-solid).

travelling across the boundary, where there can be some refraction, for example, based on changing sound velocity. It should be noted that there may also be an acoustic impedance mismatch between the transducer and Medium 1. This is where matching layers become important for some classes of ultrasonic transducer. If acoustic impedances are too different in magnitude from one interfacing medium to another, then there is a risk of significant energy loss in the system, including reflections of ultrasonic waves back towards a source. Acoustic impedance is strongly dependent on the density of the material and the velocity of the sound. It is commonly measured in terms of Pascal-second per cubic metre ($Pa.\ s/m^3$), and is notable for its analogy, in electromechanical terms, to electrical impedance. The design of ultrasonic transducers should account for impedance, both acoustic and electrical as required, and this is discussed in greater detail in Chapter 3.

1.5   EMERGING TECHNOLOGY DEMANDS

This introductory chapter has principally consisted of a broad overview of the field of ultrasonics and the key transducer configurations in use today across disciplines in both medical and industrial fields. Up to this stage, one of the core themes of this book in realising adaptive ultrasonic transducers relates to the integration of advanced materials in current configurations of ultrasonic devices. However, there are more expansive demands from industry for a wider scope in how adaptive technologies can be realised. Materials science sits on one side of this approach, but there are a range of systematic signal processing developments which can be made, for example, in how we shape and control ultrasound waves. Therefore, the later chapters of this book will aim to provide informative overviews of several key areas across materials, transducer design, dynamics, and signal processing.

Emerging technology demands can be thought of in terms of three key parameters as related to ultrasonic transducers, but which can be more broadly expressed as relevant to different applications. The first technology demand is that associated with *tuneability*, which informs us on how readily a property or characteristic feature of an ultrasonic transducer can be controlled. The most common and obvious property of interest for an ultrasonic transducer is related to frequency. The operation of an ultrasonic transducer in a specific resonant mode is generally tuned to a frequency of interest, but often this tuned frequency is fixed for many configurations of transducer, where it is not able to be adjusted. The frequency at which an ultrasonic transducer operates is important, as this might be a frequency distinct from another source in a system, or it may have some importance depending on the nature of the material or substrate involved. There are examples of frequency dependence and the need for tuneability in performance, for example, in the evident frequency dependence of how some materials can be penetrated more readily than others. This has been demonstrated in ultrasonic surgery, where it has been reported there are optimal dynamic conditions for devices penetrating softer tissues, such as muscle, compared to harder materials like bone (Lucas et al., 2023; Schafer and Cleary, 2023). Although much further research is required in these areas, the need for tuneable ultrasonic technologies is apparent. However, it should be noted that tuneability is not only restricted to frequency but instead affects a wide range of properties associated with ultrasonic transducers and their constituent materials, such as elastic modulus of a material, or a vibration amplitude of a transducer. Much of the tuneability aspect of the philosophy associated with adaptive ultrasonics is expanded upon in Chapter 4, concerning several materials challenges associated with the key advanced material classes of shape memory alloys and metamaterials.

The second technology demand is centred on the idea of *resilience*, and this is a general characteristic of an ultrasonic device which is becoming increasingly important. Here, resilience refers to a recoverable or stable dynamic response of an ultrasonic transducer across a range of changing environmental conditions. There have been significant developments in recent years to develop new methods of fabricating ultrasonic transducers resilient to a wide range of environmental conditions, for example, elevated levels of pressure or temperature. If we take the former condition as an illustration, only in recent years has the conventional and commercially available 40 kHz flexural ultrasonic transducer been redesigned and upgraded into a transmitting or receiving device resilient into the hundreds of bar (Feeney et al., 2019b, 2020, 2023; Somerset et al., 2021, 2022), far exceeding the ambient 1 bar pressure limit for which it was originally designed. This was achieved through a relatively complex and intricate adaption of the conventional transducer architecture, where the research achievements were demonstrated in the laboratory and with industry support. This is one example of the evident need for resilient technologies and where alternative solutions to this challenge would have significant potential across a variety of applications.

In general, there is a real opportunity presented by the design philosophy associated with adaptive ultrasonics, for delivering on the resilience demand. If we consider the case of the flexural ultrasonic transducer for resilience and operation in environments of elevated pressure, the status of this technology is that whilst it works well as a concept, it is highly complex to deliver on an industrial scale. It requires simplification and to be made more cost-effective, but whilst the evidence is there that it works, it is also key that the real need for such resilience exists. One key advantage of an advanced material such as the shape memory alloy Nitinol is that it exhibits a significant capacity for shape recovery and (remote or self) healing (Meyer, Jr. and Newnham, 2000). In theory, a configuration of flexural ultrasonic transducer fabricated using Nitinol could be engineered to remotely repair its own structure in the event of damage or deformation incurred through, for example, excessive levels of pressure on the transducer's vibrating membrane or plate. In another example related to this concept of remote operation, resilience via advanced, or shape memory, materials has been noted as a key factor for new technologies for space exploration (Harkness et al., 2014; Costanza and Tata, 2020; Malik et al., 2021). One advantage has been noted as the capacity for mass reduction, important for space-based and airborne operations (including drones) but also because, providing they are correctly designed and integrated, advanced materials like Nitinol can in theory actuate in different ways and replace the complex and comparatively heavy electronics which would typically be required to undertake similar routines. In principle, this would give a degree of resilience to an application, beyond what we might consider in terms of environments of hostile temperature, pressure, humidity, radioactivity, or corrosiveness, for example, in relation to what an ultrasonic transducer would normally be expected to withstand. The resilience demand is addressed in greater detail as part of Chapter 4, in the discussions on the challenges of materials science to deliver adaptive ultrasonic devices.

The third technology demand is centred on *intelligence*, and it is in some ways the most cutting edge in terms of its status of development in ultrasonic devices. Through the (relatively recent) integration of advanced materials in ultrasonic devices, for example, those where the physical responses can be trained, or where their properties must be designed since they are not found in nature, as in the case of metamaterials, we are uncovering a wide range of new capabilities with the potential to revolutionise ultrasonic processes. With the advent of such exciting new capabilities, there are new pathways towards intelligent technologies emerging, in part responding to the need for hardware and software solutions which are highly interconnected and responsive to the needs of the application or the end user. Given these demands on new technologies and the advent of more complex and pervasive artificial intelligence algorithms in society, we must give some attention to how the boundaries of ultrasonics and artificial intelligence will blur together and trigger a new subset of adaptive ultrasonic technologies. For example, we must understand how artificial intelligence could be used to program or configure the response condition of an ultrasonic device, or how it may potentially optimally select to interact with its environment. This aspect of the book is covered in detail in Chapter 5, alongside adaptive ultrasonic approaches to beam shaping and beamforming, with a primary focus on ultrasound imaging.

The technology demands of tuneability, resilience, and intelligence, as described in this chapter, are designated in this book as the three pillars of adaptive ultrasonics. They have been derived from technology demands and an element of projection into the future, but they are a useful framework from which to build up the principles of adaptive ultrasonics in this book and beyond. The general underlying philosophy of adaptive ultrasonics is not just restricted to the use of novel materials, principally shape memory alloys and metamaterials. It transcends singular or stand-alone disciplines and requires significant multidisciplinary developments across fields including materials science, mechanics, electronics, mathematics, and computing science. It is this interconnected approach towards the three pillars of tuneability, resilience, and intelligence that this book aims to explore and demonstrate.

1.6 SUMMARY

This chapter has introduced several of the key foundational principles of ultrasonics, building towards emerging technology demands for adaptive ultrasonics into the future, and it has clarified the motivation for this field of research and for the content of this book. As a broad summary of the content, this book will provide insights into contemporary ultrasonic transducer design and manufacture, an overview of advanced materials which are posited to create an impactful change in the field of ultrasonics in the coming years, and it will introduce potential applications of interest for adaptive ultrasonics, with relevant experimental case study data in support. From this point onwards, the content of this book will provide a useful starting point for any engineer from the graduate level and beyond, with insights on practical approaches to the design, manufacture, and operation of ultrasonic transducers towards adaptive ultrasonic devices.

2 The Evolution of Adaptive Ultrasonics

2.1 ULTRASOUND'S EARLY YEARS

The path that the field of ultrasonics has followed, especially since the 19th century, has been well-documented by several authors in their respective authoritative texts, including Gallego-Juárez et al. (2023a) and Ensminger and Bond (2024). However, it would be remiss not to highlight some of the key developments from a historical perspective which have led to where the discipline is today, particularly with the emergence of new themes and capabilities, such as adaptive control. The historical overview contained in this chapter is not all encompassing, and the reader is encouraged to read wider on key developments where relevant. Some of the content of this chapter is an expansion of the detailed historical account presented in Graff (1981), with supporting material where relevant from Feeney (2014). Further advances in the fields of adaptive beamforming and signal processing, and research developments related to artificial intelligence and machine learning, are provided in Chapter 5 for reference and wider reading.

The study of sound as a phenomenon can be traced back to Pythagoras, where it has been reported that he became intrigued by the impact of blacksmith tools on anvils (Caleon and Subramaniam, 2007). Very quickly, it was recognised that there were differences in terms of consonant and dissonant sounds, and that these had a distinct relationship with physical properties including weight. This significant observation led, at least in part, to a more rounded physical understanding of how different sounds, or notes, could be created using basic device configurations. One such device involved string, and it could be demonstrated how sound can be tuned depending on the selected length ratio. One of the major underlying achievements at this time was that there was a mathematical basis for sound, and it also showed that sound was a tuneable property, a concept that is fundamental to the core themes of this book.

Sometime later, it was also reported that Aristotle made efforts to understand sound in terms of its propagation characteristics, where he attempted to interpret sound propagation as a phenomenon which relied on the vibration of air (Caleon and Subramaniam, 2007). It is remarkable that even in these comparatively early times, there was a level of learning and ingenuity to make scientific developments we still recognise today, and conspicuously, progress we would recognise as important. History is not always well-documented, and it is arguable that most of the important developments in sound propagation leading up to how we understand ultrasound today would emerge much later, around the 16th century and beyond. Galileo was recorded as noting how sound could be produced in water

(Caleon and Subramaniam, 2007), and only a little later in the 17th century, Mersenne was able to demonstrate principles around the detectability of sounds and that there are boundaries or limits to what may be possible to pick up. Concepts of pitch and frequency began to emerge in greater detail and understanding, and it was from here that the scientific community started to think about frequency in more detail. One of the main drivers of this was Hooke, who notably demonstrated how it was possible to produce devices with varying frequencies (Caleon and Subramaniam, 2007), where the relationship between frequency and pitch could be understood. The device which Hooke developed was, in its most basic form, a wheel with a series of teeth which would impact a metallic component at different speeds, thereby creating musical sounds at different pitches. In the contexts of this book, this shows a very early and important drive towards adaptive or frequency tuneable technology. Even in these early years, Hooke postulated how sound could be used in medical diagnosis, when he predicted that sound could potentially be used to understand the motions or movements of the internal structures of bodies, either animal, human, or machine (Szabo, 2004). This extremely forward-thinking prediction formed a significant foundation for the entire field of ultrasound imaging, and showed early interest in driving forward scientific developments in how sound is understood and used, into new areas. If we move forward several decades to the 19th century, a concise relationship for the velocity of sound in air was determined by Laplace (Mason, 1976), shown by Equation (2.1), extracted from Wong (1986).

$$c = \sqrt{\frac{\gamma P}{\rho}} \qquad (2.1)$$

This relationship was an extension of key developments from the research of Newton in the 17th century (Caleon and Subramaniam, 2007), where Laplace correctly identified that temperature has an influence on sound velocity. This is where the relationship put forward by Newton was further developed, and this is shown via Equation (2.1) as the Newton–Laplace Equation, where γ is the thermal component, or specific heat ratio, and must be a product of the pressure P. It is customary today to note that the numerator equates to a K parameter, or a bulk elastic modulus. Put simply, if one knows the bulk elastic modulus of a material and its density ρ, then it will be possible to readily calculate the velocity of sound through that material. The nature of sound propagation in fluids was later followed by Poisson's research into how sound travels through solids (Mason, 1976).

With the concept of sound velocity and its importance in the understanding of sound wave propagation in different media, progress was then made in finding different ways of quantifying sound and measuring its characteristics. A notable step forward in the measurement of sound was made in the early 19th century by Colladon and Sturm. The basis for this was that the propagation of sound could be quantified via its velocity in a fluid, where the chosen fluid in this case was elected to be freshwater (Sherman and Butler, 2007). For obvious reasons, there was no ultrasonic or electroacoustic transducer to hand at this time, and so it was not possible to use one of these approaches to generate an excitation signal, as we might do so today. Instead, an impulse type of signal was created via the use of a bell submerged in the water. This bell could then be struck to generate a ringing sound which constituted the transmission or generation signal, which could then be collected by a suitable device. In this case, a metallic tube was submerged in water some distance from the bell, which in this case can be thought of as an early form of matched acoustic device. After using a light flash as the indicator of the signal being triggered at the bell, the time taken for the sound waves to travel over several kilometres was monitored, where the signal was collected via the metallic tube. Using the time taken for the signal to be collected from the light flash, and the known distance, a reliable measurement of sound velocity was acquired in the fluid of interest. It was later remarked by notable pioneers in the field, including Lord Rayleigh, the impressive level of accuracy that this method yielded, considering the technical instruments available at the time.

The overview of the key developments in how sound was interpreted and investigated over many years has thus far predominantly centred on thinking about sound waves in terms of what humans are able to hear. For a long time, this was the anchor point for how sound phenomena could be rationalised and understood. In the 18th century, an Italian priest and physiologist, Lazzaro Spallanzani, made arguably the first real foray into ultrasound as a concept, and the principle of echolocation (Goldman, 1962). Fascinated by how bats were able to navigate obstacles in the dark in their hunt for food at night, and to detect their prey, he hypothesised that these animals do not use their eyes to avoid collisions with objects and to track and secure their prey, but rather they rely on another mechanism which for some time remained unclear. In time, the collective efforts of Lazzaro Spallanzani and Louis Jurine eventually led them to the conclusion that it was their hearing which was central to their ability to navigate in the darkness, and not their eyesight or any other natural characteristic. Through navigation via sound, or ultrasound, waves (Cracknell, 1980), it was shown how light did not need to be a primary mechanism for how some animals could thrive, and that there was potentially a completely different world in existence which did not have the nominal bounds of the 20 Hz–20 kHz experienced by humans. At this time, the concept of ultrasound was controversial and not widely accepted, and this was despite the extensive experimentation and documentation from these notable pioneers like Spallanzani and Jurine. Nevertheless, even if some ideas take time to be accepted, research and the emergence of new knowledge continues, and towards the end of the 19th century came the famous Galton whistle, perhaps better known today as a dog whistle.

The motivation for Galton's whistle came from the eponymous creator of the device, the Victorian polymath Francis Galton. He was interested in understanding the limits of hearing in different animals, including humans, and was eventually able to demonstrate that for humans, the upper limit could be anywhere in the range from 10 kHz to 18 kHz (Richardson, 1962). Since then, this upper limit has been demonstrated more accurately. In any case, using the Galton whistle, it was unequivocally demonstrated that there were sounds certain animals could detect, which humans could not, and this would therefore fit well with the earlier observations of sound being a viable navigation mechanism for bats that is not detectable for humans. It should be noted that an upper limit estimate of 18 kHz is extremely accurate, especially at that time, because whilst the nominal upper limit of human hearing is regarded as approximately 20 kHz, this upper limit is known to reduce with age, or some illnesses or physical conditions. This upper limit can therefore reduce by a few kHz, for example, down to 17 kHz in some cases. It is more common for the young to exhibit the capability to detect sound up to around 20 kHz.

The major challenge towards the end of the 19th century was how reliable and robust generation and detection of ultrasound could be achieved, beyond somewhat unpredictable or difficult to control approaches such as impulse strikes. This was considering the greater understanding of higher frequency sound waves than ever before, harnessing the concept that there is an entire sound regime not naturally detectable by humans. As an important scientific step forward, which would transform the field of ultrasonics and propel it to a globally significant discipline for industry and medicine, the piezoelectric effect was discovered in 1880 by the Curie brothers, Pierre and Jacques (Richardson, 1962). In general terms, it was observed that a certain class of crystals could produce positive and negative charges which were in proportion to the pressure exerted in specific axes, thereby demonstrating a concrete link between the electric potential generated across the material, and the applied mechanical strain. A short while later, it was then found that this characteristic behaviour could be reversed, where an electric potential across the material could be produced in response to the application of a mechanical force, in this way being a commutative effect. Whilst this research was being undertaken, the mathematician Gabriel Lippmann was developing a thermodynamic understanding of the piezoelectric effect, and he was hence able to correctly predict the existence of this reverse piezoelectric effect (Blitz, 1963).

The real advantage of the piezoelectric effect and its commutative reverse is that it opened the door for achieving the production of highly complex and effective propagation and reception systems, where generator devices could be engineered to propagate ultrasound towards a dedicated receiver, or ultrasound detector. Without the ability to capture ultrasound, but instead only produce it, we would not be where we are today in terms of our measurement capabilities. The underlying principles associated with the generation and detection of ultrasound formed the basis of a rapid technological advancement in the 20th century, and with that would come significant efforts to control, manipulate, and tune various properties and characteristics of ultrasonic devices that would both take ultrasound technologies into new and unforeseen applications and become the adaptive technologies we are most familiar with today.

2.2 EXPANSION TO NEW APPLICATIONS

The earliest and arguably the most well-known attempt to utilise the principles of ultrasonic transduction was by one of Pierre Curie's doctoral students, the eminent French physicist and pioneer in his own right, Paul Langevin. One of the primary motivations for bringing ultrasonic transduction to the forefront of scientific development was the Titanic disaster in 1912, where new methods of detecting obstacles, such as icebergs, in the seas were urgently needed. This was also compounded by the fact that because of such maritime disasters, there were changes to the legal requirements for the safety features that vessels had to accommodate. At this time, the principles of ultrasound and transduction were reasonably well understood, and several proposed the idea of making use of ultrasound as a detection mechanism at sea, including the physicist and mathematician Lewis Fry Richardson (Cracknell, 1980). It was here that Paul Langevin innovated a novel configuration of a hydrophone making use of his doctoral adviser's piezoelectric effect, but it was several years later before it was able to be used in practice.

It was not just the changes to maritime law or disasters involving passenger vessels which accelerated the need for reliable ultrasonic detection at sea. Submarines were used in World War I in combat and thus along with that came the need to ensure that enemy submarines could be detected, and hence avoided, over great distances. There also needed to be sufficient levels of accuracy and precision, and these were some of the demands on the required hydrophone system. In general terms, Paul Langevin conceived of quartz-based solution, since quartz is a naturally occurring material which exhibits piezoelectric properties. By stimulating sections of quartz crystal into a resonant condition with a tuned electrical circuit, ultrasound waves could be generated (Richardson, 1962). It was especially significant that around this time, quartz was noted for its abilities to produce both generators and detectors of acceptable performance, giving rise to hydrophones utilising the piezoelectric effect as a viable mechanism for ultrasonic transduction in water. A little later, Paul Langevin led the development of an adapted form of the quartz hydrophone, which incorporated steel plates to sandwich the piezoelectric quartz, with the resonance governed by the thickness of the device assembly (Graff, 1981). This eventually aided the achievable energy output of the transducer and undoubtedly improved the robustness of the configurations. Of course, it shouldn't be underestimated how much research and development activity was taking place around this time, but Paul Langevin was certainly a key central figure, and the eponymous Langevin transducer today is testament to his significant contributions.

Although these early ultrasonic transducers made their mark on the scientific landscape, they were not fully realised until after World War I concluded, which was the origin of their core purpose. The Swiss physicist Constantin Chilowsky was a key collaborator of Paul Langevin, and by the time the conflict ended, they were able to demonstrate echolocation and ranging using a piezoelectric transducer to detect submarines in seawater. What was noticeable was the evident power of ultrasonic transduction and how it might transform technologies into the 20th century. In the experimentation of these new ultrasonic transducers, the destructive impact of their operation on fish was noted, and efforts were then put in place to understand the safety implications of ultrasonic technology (Szabo, 2004). What follows here is a selection of highlights on the emergence of the ultrasound scanner for monitoring pregnancy in the middle 20th century, arguably the most familiar and popular application of ultrasound up to the present day.

In the 1930s, Sergei Sokolov successfully proposed how ultrasound could be used to detect flaws or features in a variety of structures, giving rise to a completely new field of nondestructive testing and evaluation. The implications of this were significant, because it established the concept of monitoring and evaluation for identifying and isolating structural failure zones (Blitz, 1963). This concept was given further credibility through the advent of the pulse technique. Around the time of World War II, materials science was fully embedded in the research and development of ultrasonic transducers, where novel forms of piezoelectric material were under consideration. Materials science remains a critically important pillar of ultrasonic transducer research to this day, and the work begun at this time was instrumental in the acceleration of our ultrasonic capabilities through the 20th century. It was customary to integrate crystals of Rochelle salt, a piezoelectric material, into configurations of ultrasonic transducer, or alternatively a magnetostrictive approach using nickel was occasionally used. However, the Langevin transducer would typically rely on the piezoelectric effect, and other materials would be trialled, including crystals of ammonium dihydrogen phosphate, which were found to exhibit superior properties in terms of their performance in an ultrasonic transducer, compared to Rochelle salt. After World War II, technology development not only continued but also rapidly accelerated, because the barriers that existed between nations formerly in conflict were no longer there. Many developments in transduction were undertaken in Japan but did not cross international boundaries until World War II had ended. One example of technological advancement was that of the pulse technique, so important for both nondestructive testing and evaluation, and the medical diagnostics field, which relied on the innovations in radar to achieve practical implementation (Blitz, 1963). In terms of materials research, ferroelectrics, which exhibit natural and reversible polarisation, and can be found in compositions such as barium titanate, were integrated into ultrasonic transducers in the 1940s (Mason, 1976). Interestingly, the United States and USSR were developing barium titanate simultaneously and independently. However, it was discovered that several material properties associated with this type of ceramic, notably the dielectric and piezoelectric constants, and its Young's modulus, exhibited significant deviations associated with their phase microstructures. This can result in mechanical and dielectric losses, and changes in permittivity, and the material is typically doped with elements such as niobium to overcome such limitations (Cheng et al., 1996). To address this, different piezoelectric ceramic types were used, where the most popular and successful were lead zirconate and lead titanate. The primary advantage of these novel compositions was that they exhibited significantly higher dielectric and piezoelectric properties, which would thus markedly enhance the dynamic performance of an ultrasonic transducer composed of them. It is notable that today, variants of piezoelectric ceramics consisting of lead zirconate titanate (PZT) remain the most popular and widely used compositions across medicine and industry.

One of the drives of piezoelectric ultrasonic technology into new application spaces was the identification and development of technologies able to operate at higher amplitudes. Wood and Loomis are among the first to recognise that through the application of higher power ultrasonics, a fog-like effect, or the generation of cavities, could be produced in liquid (Mason, 1976). These higher intensities were clearly producing a different effect in the medium compared to ultrasound waves at lower amplitude levels. High-power ultrasound opened the door to the disciplines of ultrasonic cleaning (which is as popular as ever today), and liquid food degassing (Cracknell, 1980; Knorr et al., 2004). It is notable that in the middle of the 20th century, new applications were continually being sought from a fundamental technology base, and it is arguable that adaptive or tuneable technologies have been in demand significantly longer than one would perhaps appreciate. Beyond any industrial application of ultrasound, there was growing activity in investigating how ultrasound could be reliably applied to the field of medicine (Szabo, 2004). Again, around the middle of the 20th century, various uses of ultrasound waves to characterise aspects of the human body were proposed, including general imaging of parts of the body. The central enabler of these early developments was the pulse-echo method, where a signal is generated by a transducer and the evidence of the echo is then monitored, usually by the same device. The transfer of ultrasound waves from a generator device to be collected elsewhere, for example, a secondary matched transducer, is normally regarded as the pitch-catch method.

The interest in harnessing ultrasound for medical applications rapidly gathered pace, and this time in the development of ultrasound technology experienced a significant jump in application and complexity. It quickly became evident that should ultrasound be taken seriously for medical procedures, there would need to be a minimum level of control in how a device could propagate and detect ultrasound waves in the human body, and the achievable level of resolution of the generated ultrasound images. This is where Karl Dussik stepped in, an Austrian psychiatrist who was particularly interested in understanding and diagnosing conditions related to brain tumours (Szabo, 2004). It is an interesting coincidence that he happened to come across the exciting opportunities afforded by ultrasound waves for the detection of objects at sea and in the water in general, and he quickly took it upon himself to investigate how ultrasound could be used non-invasively for his medical purposes. He was able to create the first ultrasound images of the human brain to detect tumours, which were referred to as hyperphonograms, setting the foundation for important research that continues to this day. Karl Dussik correctly envisaged that, if applied in the right way, ultrasound could be used to differentiate between different structures or tissues in the human body. Following these developments, other scientists and engineers took the technology forward in different ways and for a range of diagnostic procedures, including diagnosing gall stones (Szabo, 2004), and the understanding that ultrasound waves can propagate through a variety of materials at different velocities, thus providing a mechanism to differentiate between different organs, tissues, or cavities in a human body (Szabo, 2004). This research activity at the time also famously led to the application of ultrasound in obstetrics in the 1950s by Professor Ian Donald (Donald and Abdulla, 1967; Donald, 1974), who was Professor of Regius Midwifery at the

University of Glasgow at the time. He became particularly interested in what ultrasound could do for medicine after learning of its capabilities for nondestructive testing and evaluation in various industrial capacities. In addition to the critical developments for monitoring pregnancy which steadily developed over the subsequent years, Professor Donald and colleagues made significant contributions to cyst profiling and sonar in general (Donald, 1974).

It is arguable that up until the middle of the 20th century, there was limited focus on what we would understand today as power ultrasonics. This contrasts with the progress made in nondestructive testing and evaluation, and ultrasound imaging. It is perhaps understandable, given the technological constraints at the time. Around the 1950s, research came to prominence with W.P. Mason reporting the fabrication and trial of a transducer with a horn attached, with the specific objective of amplifying the velocity at the resonance of a piezoelectric ultrasonic transducer (Graff, 1981). Mason was responsible for proposing an electromechanical model of the piezoelectric ultrasonic transducer, and he used all relevant research developments to henceforth deliver the ultrasonic horn as a key constituent of an ultrasonic transducer. This research and development, in general, allowed these devices to be used at higher levels of amplitude and in the interaction with a range of solid materials (Graff, 1981), not previously possible. The embodiment of the classical form of the Langevin transducer began to emerge from this point, and the next major contribution was the implementation of pre-stress, or pre-loading, on the central bolt of the Langevin transducer. This is effectively a mechanical bias which allows piezoelectric materials that are often inherently brittle in nature or liable to fracture under high loads to be driven at higher amplitudes than would otherwise be possible (DeAngelis et al., 2015; Mathieson and DeAngelis, 2015; Li et al., 2019). Pre-stressing and the tuned resonant horn were two notable developments of the 1950s and 1960s which opened a plethora of viable new applications and capabilities for ultrasonic transducers. Power ultrasonics could then branch out in a significant way to a variety of fabrication technologies, including the joining, welding, and forming of a variety of metallic and non-metallic (such as plastic) materials, revolutionising manufacturing capabilities.

The expansion of ultrasonic devices to new applications, and where we will progress towards adaptive ultrasonic technologies, should also include some level of commentary on ultrasonic surgery. This stemmed from the notable progress made in the middle of the 20th century in enabling piezoelectric ultrasonic transducers to be operated at higher amplitudes, and the demonstration of tuned horns which in part enabled the amplitude to be controlled in some capacity, but also ensured the area of the device interacting with a target tissue would be as small as possible. Ultrasonic transducers for surgery were originally targeted for dentistry in the 1950s (Graff, 1981), and principally excited through the magnetostrictive effect, rather than by using piezoelectric materials which are more popular today. The developments in these ultrasonic transducers eventually led to the ability to treat a range of conditions, such as cataracts (Lucas et al., 2023), and towards the latter half of the 20th century there was a surge in activity addressing how these ultrasonic transducers could be used to operate on soft muscular or hard bone tissues alike (Lucas et al., 2023; Schafer and Cleary, 2023). New opportunities emerged in how such ultrasonic surgical devices could be used to not only cut or penetrate biological tissues but also preserve as much of the live tissue as possible, reducing tissue necrosis, and ensuring that biological materials can be cut and operated on with minimal force and highest precision possible (Amaral and Chrostek, 1997). This research continues to this day, and it forms a key part of the underlying motivation towards the development and deployment of adaptive ultrasonic devices.

In general, ultrasonically assisted phenomena, such as cavitation, wave propagation, high-amplitude cyclic vibration of tuned horns, and the generation of standing waves, feature in the following (non-exhaustive) list of applications that have expanded the field of ultrasonics rapidly from the 20th century to the present day:

1. Cleaning
2. Compaction and consolidation of particles
3. Drilling
4. Flow measurement
5. Imaging
6. Metrology
7. Micromanipulation of particles in fluids
8. Mid-air haptics via standing waves
9. Precision surgery in medicine and dentistry
10. Robotics
11. Sensing technologies and nondestructive testing and evaluation
12. Sonar
13. Therapy
14. Welding of dissimilar materials
15. Wire bonding

The bibliography of this book contains the details of notable scientific literature sources related to each of the areas listed. There are numerous opportunities for a wide variety of modern technologies afforded by the range of ultrasonic devices we now have at our disposal. With regards to robotics, there has been significant progress made in artificial intelligence and how we can use it in ultrasound, and this will be discussed later in this book, as part of Chapter 5. With the modern capabilities in ultrasound and the range of ultrasonic devices we can devise and construct, for the most part based on the similar key scientific and engineering principles which have been central to this field for so long, there has been a steady drive towards tuneability and adaptive control of various associated phenomena. The remainder of this chapter will provide an overview of some of the origins of this drive, and its importance to the field of ultrasonics in general.

2.3   PATHWAYS TO ADAPTIVE ULTRASONICS

Before we delve into the particulars of adaptive ultrasonics, it is worth directing some focus to adaptive technologies from a more general perspective, and the underlying mechanisms to what constitutes enablers of adaptive features of an engineering system. If we return to the central definition of what adaptive means in the contexts of this book, it is something that is designed to automatically adjust as a response to a change in a particular parameter, which could be (not limited to) a performance indicator or an external condition. One adaptive technology which might immediately spring to the mind of the reader is an adaptive system for helping those with additional needs, for example, speech recognition where required, for instance, in the case of a system which may respond to a sound input or to text on a page. Another example of an adaptive technology is artificial intelligence, which is a broad concept, but in some cases constitutes technology that can respond to changes in key system parameters to engage in a series of decisions or commands based on prerequisite or current information.

In part alluded to in Section 1.5, a key basis of adaptive technology is multi-functionality, or expressed in an alternative way, the property of a quantity or characteristic behaviour being tuneable such that the system functionality can be adjusted or adapted as required. If we consider the principles of tuneability, resilience, and intelligence, as central to an adaptive ultrasonic technology and detailed in Chapter 1, then we can set some boundaries on the parameters or characteristics that we would consider an adaptive ultrasonic transducer might constitute. Therefore, if we take the concept of a conventional configuration of ultrasonic transducer, we can ask ourselves – what do we need to do to convert it to an adaptive configuration? In the context of the content presented in this book, the first stage is to consider the materials we can use. The thesis is that advanced materials, predominantly those exhibiting shape memory or controllable properties related to the phase microstructures of the material, present a viable route to tuneability and resilience. Metamaterials also exhibit real opportunities as an alternative route to tuneable features of an adaptive ultrasonic device, and they are henceforth considered. The next way we can think of adaptive behaviours is through the control of the propagated ultrasound wave, for example, in a mechanism we would term beamforming. This is based in signal processing, but it must be considered as another strand of adaptive ultrasonic technologies. For these reasons, Chapter 4 addresses the opportunities afforded by advanced materials and provides advice for integrating them in ultrasonic transducers, whereas Chapter 5 details the signal processing strategies for realising adaptive beamforming and beam shaping. Together, these approaches are postulated to be viable and central to future adaptive ultrasonic technologies. The third mechanism which is treated as part of Chapter 5 is the integration of artificial intelligence in the application of ultrasonic devices and their associated systems. It is likely that the integration of artificial intelligence features and strategies in ultrasonics will become more prominent in the coming years, and so it is appropriate to include some contextual information in this book.

The most popular shape memory material, by some distance, which has found distinction in a range of engineering technologies is nickel titanium, or Nitinol. The properties of this material and its applicability to ultrasonics are explored in detail in Chapter 4, but it is important to flag at this stage that a major part of the focus on materials science reported in this book relates to Nitinol, given its global status as a particularly commercially successful shape memory alloy. In terms of a pathway to adaptive ultrasonics, prior to the 21st century there was relatively little in the published literature which demonstrated the incorporation of these materials in ultrasonic devices. Even the major developments regarding metamaterials did not come to prominence and accelerate until the late 20th century, in a major way led by Sir John Pendry. Therefore, it should be no surprise that the key developments have not come until relatively recently, in the early 21st century.

It was clearly argued in 1999 that there was a real need emerging for the number of applications in which shape memory alloys are integrated and utilised (Van Humbeeck, 1999). As a material used in engineering systems, Nitinol has experienced arguably significant attention, but only for a relatively small number of applications. These include biomedical stents (Sachdeva et al., 2001), aerospace systems including damping structures (Alzhanov et al., 2023), and micro-actuators (Hagemann et al., 2015). The main limiter for the integration of shape memory alloys into ultrasonic devices, and some related technologies, has been that they exhibit highly complex thermomechanical behaviours that have taken some time for materials scientists and engineers to understand (Stöckel, 1995; Duerig et al., 1999). There is an active research community aiming to understand various aspects of the mechanical behaviour of shape memory materials, with one potential outcome being their exploitation in ultrasonic devices, and another with a close relationship, in acoustic resonators.

Shape memory alloys such as Nitinol have commonly been utilised for their superelastic behaviour, which is the relatively high recovery possible in response to a particular level of applied stress. The superelastic effect is an underlying mechanism for biomedical stents (Robertson and Ritchie, 2008), but it is evident and arguable from the literature that the shape memory effect, the ability of a shape memory alloy to recall and restructure itself into its original set shape in response to an external trigger such as temperature, has been underused by comparison. The thermomechanical behaviours of a shape memory alloy like Nitinol will be addressed in more detail in Chapter 4, but principally both superelasticity and the shape memory effect rely on phase transformations of the material's microstructure. In Nitinol, this microstructure can either be martensitic, often with a monoclinic crystal structure, or a stiffer cubic austenite. The elastic moduli associated with these microstructures are also different, thus giving rise to the possibility that by simply changing the dominant material in an ultrasonic transducer, which is not the piezoelectric or magnetostrictive element, the resonance frequencies and dynamic characteristics of the device can be tuned, thereby delivering an adaptive ultrasonic device. In theory, this could all be achieved in principle using a variety of shape memory materials, but relatively little progress has been made, except for Nitinol and some of its variants, and therefore much of the relevant content provided in Chapter 4 focuses on this material.

A key restriction for utilising the shape memory effect in devices is that there should be precision in understanding the transformation temperatures of the material (Stöckel, 1995), or those temperatures at which the shape memory material will transition from one phase microstructure into another. This has evidently been a barrier to a pathway to adaptive ultrasonic devices in the contexts of tuneability and resilience. One could simply design an ultrasonic device where it would be possible to interchange end-effectors with ease, but this would not strictly meet the criteria of an adaptive ultrasonic device, even if it would yield a device with consistent dynamic features. There must be methods to understand how we can integrate advanced materials in ultrasonic transducers, with a coherent and comprehensive design strategy. Given that the literature is not replete with published reports in this area, only key examples of notable developments can be highlighted, and where not all are limited to being integrated with shape memory materials.

An important example of a drive towards adaptive ultrasonic technology was presented in the middle of the 1990s, in the form of a tuneable ultrasonic device consisting of Nitinol with a postulated application of imaging (Lorraine, 1995). This device was patented and demonstrated the principle of possessing a centre frequency which could be selected. A notable example of adaptive ultrasonic technology from the early 21st century was an actuator device utilising the two-way shape memory effect (Kim et al., 2008), which itself was an important development in the field as it demonstrated a viable method of utilising two-way shape memory, or making use of more than one set shape of the material which could be activated by an external trigger. A major advantage of this approach which was noted was that a residual strain in the material would not need to be created. A report demonstrating the triple shape memory effect was published only a short time later in 2012, exploiting the rhombohedral R-phase, which is a third form of microstructure which can appear in the phase transition profile of a shape memory material such as Nitinol.

We can momentarily consider the drive towards multi-frequency and tuneable dynamic devices and systems, not limited to ultrasonics. The advantages of technology of this nature have been known or remarked upon for several decades, but it is arguable that the technological capabilities have not been available until more recent times to deliver on some of these ambitious propositions. There was progress towards a multi-frequency transducer in the 1970s making use of low- and high-frequency elements in an array configuration, permitting control (Madison and Frey, 1976). In another important contribution to underwater transducers, a device configuration was proposed in the 1980s for a dual-mode broad bandwidth transducer (Lindberg, 1983), and around the same time other reports of dual-mode transducers emerged. One of these was in effect a composite device which allowed resonance at different, separate, frequencies (Thompson, 1986a), and the second was a form of longitudinal transducer which was broadband in nature (Thompson, 1986b). Another key development came a little later in the middle of the 1990s, with the development of a dual-frequency device which consisted of Tonpilz-like elements, and which was demonstrated to be capable of operating at two separate and distinct resonance frequencies, either at the same time or independently (Stearns et al., 1996). Around the same time, James Tressler and Robert Newnham proposed, what they designated as, doubly resonant cymbal transducers (Tressler and Newnham, 1997), with a significant advantage being the ability to operate at extremely wide bandwidths, in relative terms. The use of a cymbal transducer in this research was particularly interesting and intuitive, because it is a relatively straightforward configuration of device which lends itself favourably for integrating advanced materials, and particularly those which are difficult to machine or mechanically process. Robert Newnham would remain as one of the key contributors to this type of technology, including in the incorporation of Nitinol shape memory alloy into the ultrasonic cymbal transducer in 2000 with Richard Meyer, Jr. (Meyer, Jr. and Newnham, 2000).

In the early 2000s, research activity in this area accelerated, with a notable example of a multi-purpose ultrasonic slotted array device reported in 2003 (Mauchamp and Flesch, 2003), postulated for use in medical applications. There was also the development of a sonar device capable of resonance at three individual frequencies, based on the device assembly component being actuated or vibrated (Porzio, 2009). Again, a major motivation of multiple frequency ultrasonic transducers for an application such as underwater communications or sonar is to significantly broaden the bandwidth. More recently, there has been interest in understanding how to exploit various resonant modes of the flexural ultrasonic transducer, based on the axisymmetric plate modes of its vibrating plate or membrane (Dixon et al., 2017; Feeney et al., 2018a, 2018b). One advantage that this would afford, would be the ability to control the resonance frequencies of the transducer in an application such as flow measurement. This might be important if there is a source in the proximity of the transducer which is close in terms of its frequency, but which should be avoided. Current commercial configurations of flexural ultrasonic transducer possess a resonant mode with optimally high vibration amplitude, in comparative terms, where switching to a vibration mode of a different order may significantly reduce the vibration amplitude. It is possible that by using a phase-transforming shape memory alloy as the transducer plate or membrane material, this limitation could be overcome.

The application of smart materials, such as shape memory alloys, in the design and fabrication of transducers has been recorded as early as 1992 (Newnham, 1992), where around the same time there were developments in the depositing films of piezoelectric lead zirconate titanate in the production of composites exhibiting smart behaviours, in the interests of optimising the design of transducers with tuneable characteristics (Chen et al., 1992). There are reports of continued research and development activity concerning tuneable devices in the 1990s, including the patenting of a tuneable frequency Langevin ultrasonic transducer, by Newnham et al., where the resonant frequency could be tuned and controlled via the compressive stress applied to the composite (Newnham et al., 1992; Meyer, Jr. and Newnham, 2000). Further research in the mid-1990s then investigated alternative sandwich configurations (Alwi et al., 1996), demonstrating a continued appetite for adaptive transducer solutions. It is evident that there was an emerging pathway towards the integration of alternative materials in ultrasonic transducers at this time of ultrasonic transducer development, and by the end of the 1990s there were reports of adaptive tuned vibration absorbers (also referred to as ATVAs) fabricated using Nitinol shape memory alloy (Williams et al., 1999). At the start of the 21st century, in 2000, magnetostrictive materials, principally Terfenol-D, were integrated into devices such as tuneable vibration absorbers (Flatau et al., 2000). This was principally undertaken to control the resonant frequency of the system, and this culminated in a major milestone that same year with the publication of an adaptive frequency cymbal transducer fabricated using Nitinol, by Meyer, Jr. and Newnham, as referred to previously (Meyer, Jr. and Newnham, 2000). A major outcome of this research, in addition to the demonstration that resonant frequency could be tuned and expediently controlled, was that there was a capacity for the remote repair of a transducer by exploiting the shape memory effect of the Nitinol. The implications of this are potentially significant, as it means that a damaged ultrasonic transducer can be remotely healed through temperature stimuli. Depending on the composition and processing conditions used to fabricate the shape memory alloy, the necessary temperature gradient could be modest and is one area of research under active investigation.

In the 20 years leading up to the writing and production of this book, there have been several, albeit sporadic and disparate, investigations relating to the integration of shape memory materials into sensors and actuators, and very little in terms of those for ultrasonics applications. It is evident that the challenges surrounding controlling and understanding the material properties of shape memory alloys have been barriers to exploitation and wider application, but there have nevertheless been important steps forward. Shape memory alloy actuators have been the subject of intellectual property including patenting at least since 2000, where a notable example is a patented actuator from 2002 able to transform its microstructure, and hence its characteristic response, between two different microstructural states (Ashurst, 2002). Here, these two states have been noted as a stiff cubic austenite and a conventionally more compliant rhombohedral microstructure, usually known as the R-phase. These microstructures are discussed in more detail later in this book. The major advantage of utilising a phase microstructure like the R-phase is that the temperature gradient required to transform can be much narrower than if one was to transform between complete martensite and austenite. However, the magnitude of the change to key material properties may not be as high, and this would have implications for the tuneability capacity and performance of the device.

Further to this, a patent based on research into the design and fabrication of tuneable ultrasonic devices followed a year later (Fenton et al., 2003), demonstrating a clear need for ultrasonic devices exhibiting tuneable dynamic properties. The tuneability in this case was achieved based on adjusting the geometry of the device, but it was an important development because it showed the applicability for medical and surgical procedures. In another application of Nitinol shape memory alloy, it was demonstrated in the mid-2000s that stress-biasing of a flextensional cymbal transducer could be achieved using the material which led to notable performance enhancements (Narayanan et al., 2007). By making use of features such as the shape memory effect, it was demonstrated that it was possible to optimise the performance of transducers. Around the same time, there were active conversations on the use of shape memory alloys for actuators and other oscillating and vibratory systems (Mertmann and Vergani, 2008), although it should be noted that shape memory alloys remained particularly popular in their wire forms. This is because of their relative ease of manufacture and their subsequent moderately low cost. Reports in the mid-2000s also began to explore energy harvesting capabilities, and the various approaches in device design to enable innovative new technology solutions. Such reports included the proposition of a tuneable piezoelectric-based transducer system with adjustable structural stiffness (Peters et al., 2008; Li et al., 2011), demonstrating the drive towards different solutions to enable adaptive devices in the future. Research into energy harvesting has continued to grow ever since, and it reflects the momentum of science and engineering towards sustainable technologies.

Since 2010, a significant proportion of the research focused on incorporating shape memory alloys in ultrasonic devices has been limited to Nitinol and for the flextensional cymbal transducer (Meyer, Jr. and Newnham, 2000; Feeney and Lucas, 2014, 2016, 2018; Smith et al., 2022). There have been significant efforts to introduce adaptable or adjustable features into devices, with one notable example being the incorporation of Nitinol into a surgical needle prototype to control the location of the surgical needle tip (Datla et al., 2012), thereby promoting precision in surgical procedures. Another major strand of research activity has focused on the characterisation of shape memory materials using ultrasound approaches, for example, resonant ultrasound spectroscopy. An interesting report of steerable unidirectional wave emission using a shape memory alloy metasurface was published in 2020 (Song and Shen, 2020), showing a novel adaptive ultrasonic device in the metamaterial space. Some of these developments will be captured in greater detail as part of Chapter 4. In general, the research conducted into the cymbal transducer and the incorporation of Nitinol end-caps in the fabrication of adaptive flextensional ultrasonic transducers has laid the foundation for the development of Langevin ultrasonic transducers incorporating Nitinol, either as the end-masses in a more conventional configuration or as rings in cascaded forms of transducer (Liu et al., 2024b). The design, fabrication, and characterisation challenges associated with integrating advanced materials in ultrasonic devices are significant, both for shape memory materials and for metamaterials. These have been the principal limiters for continued development in this field, and for the success of adaptive ultrasonics as a stand-alone discipline. The key existing reports in the scientific literature have demonstrated the need for such adaptive technologies, and it is incumbent on the scientific and engineering communities to progress our understanding of materials and dynamics such that innovative new configurations of ultrasonic transducer fabricated using these materials can be delivered, to thereby enhance the wide range of medical and industrial applications available.

2.4   SUMMARY

This chapter has provided a brief but expansive overview of the evolution of ultrasonics from the late 19th century, through a selection of the relevant key technological advancements of the 20th century, to the emergence of adaptive ultrasonic technologies as a new pathway for the field of ultrasonics into the 20th century. It is evident that the early progress in ultrasonic technologies set a strong foundation for the more recent research activities which are making such impactful changes to society and industry in modern times. It is interesting to observe the drive to widen application spaces and explore new opportunities, even from the investigations conducted at the turn of the 20th century. It is arguable that technological advancements have predominantly been enabled and driven through a commercial, industrial, or military need. For example, this can be attributed to a technology such as the application of pulse-echo ultrasound in obstetrics. Although a subject matter which became apparent to Professor Ian Donald through his knowledge of industrial measurement technologies, it is evident that there was a clear need at the time for more accurate and reliable obstetrics capabilities, especially from a monitoring and diagnostic point of view. The review of the scientific and engineering literature has indicated a significant emerging trend of technologies towards intelligent, controllable, or tuneable ultrasonic devices. It is likely that a coherent progression towards such adaptive ultrasonic technologies will gather pace in the next few years, but there remain significant engineering challenges associated with dynamics and materials science to overcome.

The first two chapters of this book have aimed to provide the necessary foundation for the reader, in terms of ultrasonics and its elementary and most fundamental principles, the key transducer configurations of interest across power ultrasonics and nondestructive testing and evaluation, and the historical perspective on the emergence of adaptive ultrasonics as a discipline, within the field of ultrasonics. The next step in this narrative is to understand a selection of design and manufacture principles for ultrasonic transducers, using a range of modern approaches to device component fabrication, obtained through a combination of research and experience. It is hoped that the inclusion of this material provides the necessary context for the advice presented in Chapter 4 regarding the design and construction of an adaptive ultrasonic transducer using advanced materials, followed by commentary on alternative adaptive ultrasonic device concepts.

3 Design and Fabrication Considerations for Ultrasonic Transducers

3.1 SCOPE

A selection of the important foundational principles associated with the design and mathematical modelling of ultrasonic devices are presented in this chapter, alongside some associated practical advice for the manufacture of transducers, including suggestions for how to undertake a reliable and effective experimental characterisation. Some of the key fundamental principles associated with ultrasonics, wave propagation, and the relevant historical developments in the field were presented across Chapters 1 and 2. However, the process of designing and characterising an ultrasonic transducer can be a challenging task to undertake. It is often especially difficult to fabricate transducers with a quality sufficient for practical application, and it is not always obvious which aspects of a manufacturing process will be the most time consuming or technologically demanding. In this chapter, important design and fabrication principles are discussed for ultrasonic transducers, with examples given both from the scientific literature and from the author's experience where relevant. The discussion is generic where possible, applying to the different classes of ultrasonic transducer detailed in Chapter 1. It is hoped that through the practical advice presented in this chapter, the reader can begin to experiment with different types of ultrasonic transducer and understand where to begin in the design process. Finally, the information provided in this chapter is intended to serve as an important foundation for the discussions on advanced materials presented in Chapter 4, showing where the design of an adaptive ultrasonic device could begin.

Although the practical design and fabrication advice presented in this chapter is general, specific references are made to the Langevin transducer (arguably the most popular power ultrasonic device used today), the Class V flextensional cymbal transducer (often used in underwater applications like sonar, but more recently investigated for power ultrasonic applications), and the flexural ultrasonic transducer (common for proximity sensing). Example images of these transducers can be found in Figure 1.2 for reference. It has not been possible to provide a detailed overview of the design and construction of each configuration of transducer shown in Figure 1.2, but the three identified here are relatively common and should have some interest to the reader. Furthermore, some of the design principles discussed in this chapter can be applied to more than one transducer configuration. For example,

DOI: 10.1201/9781003324126-3

there are some similarities between the flexural ultrasonic transducer and the classical piezoelectric micromachined ultrasonic transducer, since they both incorporate flexing plates or membranes, depending on their construction parameters. There are nevertheless key differences in their design and operation which make them distinct from one another. Finally, the three configurations of ultrasonic transducer chosen for reference in this chapter are all classes of transducer that have already been incorporated with advanced materials such as shape memory alloys, or they constitute ideal configurations for doing so in the future with significant potential. For example, the vibrating plate of the flexural ultrasonic transducer is typically circular and edge-clamped; therefore, it would be relatively straightforward to manufacture this shape from Nitinol shape memory alloy in the development of an adaptive ultrasonic transducer. These principles will be further discussed in Chapter 4.

This chapter begins with an overview of the transducer configurations of relevance to the content presented, and to provide some context for the design advice. Following this, a discussion on several design aspects is presented, including those important for how to incorporate modelling and simulation prior to the planning and delivery of an experimental characterisations. An overview of active material considerations is then provided, predominantly focusing on piezoelectric material which remain popular in ultrasonics. An overview of device manufacturing is then presented, with a review of some of the key approaches used for characterising the dynamic performance of an ultrasonic transducer. The chapter concludes with a brief case study on how the given advice can be used, in the design, fabrication, and characterisation of a flexural ultrasonic transducer, in this case for air-coupled ultrasonic measurement applications.

3.2 PREPARATORY ADVICE

There are numerous important texts available from experts in the field, notably but not limited to Cheeke (2010), Rossing (2014), Matsukawa et al. (2022), Gallego-Juárez et al. (2023a), and Ensminger and Bond (2024), which report on widely accepted and verified approaches to the design of ultrasonic transducers. Specifically, key texts in the field are *Ultrasonics* by Ensminger and Bond, and *Power Ultrasonics* by Gallego-Juárez and Graff, both of which detail, in significant depth, approaches to the design of ultrasonic devices from first principles, the nature of wave propagation, and the broad range of applications for which ultrasonic devices are now designed. This chapter does not intend to replicate this material to the same extent, instead aiming to add perspectives relevant to adaptive technologies and the complexities of advanced materials to the challenge of ultrasonic transducer design, manufacture, and characterisation. This chapter is structured around illustrative examples from both contemporary power ultrasonics engineering and nondestructive testing and evaluation, to provide a necessary foundation of knowledge for devising and fabricating ultrasonic transducers with adaptive features. Examples from the disciplines of power ultrasonics and nondestructive testing and evaluation are both included, principally to contrast approaches to transducer design for each. However, there are several factors which will be addressed, including the variety of application areas and some commentary on the practical constraints on the transducers in an application environment. For example, this may include the influence of physical boundary conditions on the operation of a transducer, or the sensitivity of dynamic performance to changes in these boundary conditions. The content presented in this chapter is principally focused on piezoelectric-based transducers, given their prevalence and expansive application across different disciplines, and the research which has been conducted thus far in the field of ultrasonics towards adaptive ultrasonic transducers incorporating piezoelectric materials.

In general, there are several major steps associated with the creation of an ultrasonic transducer, from the conceptual design phase through to its manufacture and operation. It can be challenging to define a universal design process for an ultrasonic transducer, given the varied nature of the transducer configurations presented in Figure 1.2. Therefore, the first part of this chapter sets out a few key questions which can be asked in the preparation stage prior to the design of a transducer. This is important so that suitable considerations are placed on each stage of the design process, and to ensure that the designer will be able to deliver a transducer to meet the demands of a target application. Following this, instructional information is presented regarding fundamental aspects of transducer design, including material selection, finite element and mathematical modelling, fabrication, and characterisation of device dynamics.

To begin, the generalised schematics of typical configurations of ultrasonic transducer considered in this book are illustrated in Figure 3.1, encompassing the Langevin transducer commonly applied in power ultrasonics applications; the flexural ultrasonic transducer for proximity sensing and industrial process applications such as flow measurement; and the flextensional cymbal configuration, which is often called the Class V transducer in the scientific literature, traditionally employed in underwater sonar but now investigated for power ultrasonics applications (Bejarano et al., 2014, 2016).

The schematics for the transducers shown in Figure 3.1 are generalised. For example, the flextensional Class V shown in Figure 3.1(b) is conventionally circular, where a section view is shown here for clarity. An example of the complete fabricated transducer can be found in Figure 1.2(d). Furthermore, the piezoelectric ceramic stack indicated for the Langevin transducer in Figure 3.1(a) is typically composed of a series of piezoelectric ceramic rings with electrode rings, usually fabricated from copper. It is usual to find two or four piezoelectric rings in a typical commercial Langevin ultrasonic transducer assembly. The flexural ultrasonic transducer shown in Figure 3.1(c) is also circular in its commercial form, as shown in Figure 1.2(e) for reference. Finally, the epoxy resin deposited in the case of the flexural ultrasonic transducer would generally be assumed to be equal in diameter to that of the piezoelectric disc, although this is difficult to ensure in practice. The epoxy resin deposited in the fabrication of a cymbal transducer would typically be found in all interfaces between the piezoelectric ceramic disc and the end-caps, such as the flanges.

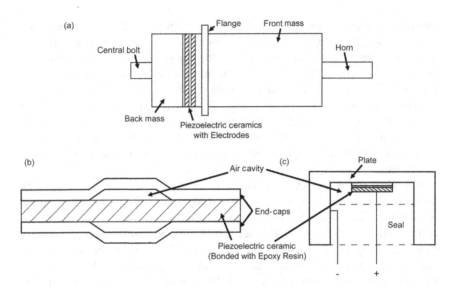

FIGURE 3.1 Generalised schematics of ultrasonic transducer configurations employed in contemporary industrial and medical ultrasonics applications, comprising (a) the Langevin, (b) the flextensional Class V (side section view), and (c) the flexural.

The scientific literature details many adapted forms of the configurations depicted in Figure 3.1, implemented across several medical and industrial applications, and as such these configurations that are included this book should provide a sufficient overview for addressing a broad range of practical ultrasound problems. For example, Langevin transducers with radiator plates attached have been widely developed and reported (Gallego-Juárez et al., 2023b), in the processing of gases or liquids, and to promote cavitation within certain media. As another example, the concept underlying the flexural ultrasonic transducer configuration has been applied in the construction of phased array devices for highly accurate and robust in-situ flow measurement (Kang et al., 2017, 2019), where multiple flexurally vibrating elements have been fabricated and tuned together to exhibit directional ultrasound wave propagation and sensing. Therefore, the transducer configurations shown in Figure 3.1 encompass a broad array of possibilities. There are several fabrication considerations which directly impact the assembly processes of all these transducers together, one example being the mechanical coupling. In general, the steps associated with the design and fabrication of an ultrasonic transducer (which is sometimes used interchangeably here with *device*) can broadly be summarised in eight steps, detailed below.

1. Define the operational parameters of the system.

The important factors to consider for ultrasonic transducer design and manufacture include the required operating frequency, desired vibration or displacement amplitude, knowledge of the boundary conditions on the transducer embedded in its application environment, the environmental conditions in which the transducer will be operating, and the excitation conditions that will be applied to the device. It is highly recommended that a straightforward list of design constraints is written, immediately at the start of any transducer design process. One may begin with the target frequency of operation, usually a resonance frequency, governed by Equation (3.1).

$$f = \frac{c}{\lambda} \qquad (3.1)$$

In this case, frequency f is a ratio of wave velocity c and the wavelength λ. The wave velocity itself is a square root of the ratio of Young's modulus E associated with the material and the density ρ, as per Equation (3.2).

$$c = \sqrt{\frac{E}{\rho}} \qquad (3.2)$$

These are generalised, and depending on the configuration and the transducer geometry, there will be variations on these relationships to accurately calculate parameters such as the wavelength.

2. Decide which class of transducer will be required.

There are many classes of ultrasonic transducer, and each comes with advantages and disadvantages. For example, if a transducer is required for a power ultrasonics application to deliver high-amplitude vibrations to a small area or volume, a Langevin-style transducer may be optimal, although they can take time to properly design and fabricate. If one wishes to avoid some of the complexities of building such a transducer, a flextensional configuration may be more suitable in theory, though there are challenges associated with controlling the epoxy resin bond layers, both in terms of geometrical dimensions and physical (mechanical) properties. The dynamics of these transducers can therefore be difficult to tailor with precision. Due consideration also needs to be given to the method by which the transducer will be driven, for example, via piezoelectric materials or if the excitation will be electromagnetically applied. Integration of these driving mechanisms have bearing on the structure and design of a transducer.

3. Develop a robust mathematical model or simulation.

It is important to use analytical, numerical, and finite element approaches sensibly and as tools to enable the development of high-quality ultrasonic devices suited to the intended application. For example, by anticipating the excitation conditions, the modal dynamics of an ultrasonic device can be configured, accounting for phenomena including nonlinear behaviours. Finite element techniques should also be used to optimise the performance of the device. In some circumstances, it is appropriate to undertake a comprehensive sensitivity analysis to determine optimal geometrical and material (property) combinations. In others, it is advantageous to undertake a mesh convergence study through finite element analysis. This allows the finite element mesh, or distribution and resolution of individual elements in the model, to be optimised so that the dynamics and stress distributions can be relied upon for different modes of vibration.

4. Compile a list of candidate materials for the transducer.

Only by following Steps 1 and 2 above can this be reliably achieved. For example, the choice of piezoelectric ceramic will be entirely dependent on factors such as the application (e.g., if the intended application is sensing, or if it is a destructive ultrasonic process), the temperature of the environment in which the device is being operated (the Curie temperature of a piezoelectric ceramic is critical, but the practical operating limit of the ceramic will likely be significantly lower than this, occasionally up to half depending on the piezoelectric material), and its operating modes.

The influence of environmental conditions on the transducer materials must also be considered in this step. For example, it should be known if the transducer will be operating in temperatures that will significantly influence the vibration characteristics of any component of the device. In the case of transducer classes like the classical Langevin, knowledge of the location of the nodal plane (where there is negligible displacement amplitude and where the transducer is commonly fixed in place for application) is important. It is key for understanding the boundary conditions on the transducer, but the material chosen for this part of a transducer must be able to withstand the expected stress levels, as indicated by the outcomes of Step 3. Therefore, often the processes followed in Steps 3 and 4 are exchanged and periodically revisited until a viable design is defined.

The next aspect to consider in this step is to understand if the chosen materials can be machined or fabricated in the desired shapes required for a transducer. This leads on to manufacturing considerations, but physical factors including the acoustic impedances of the different regions of the device, housings, or ancillary support materials in a transducer (such as a rear seal in the case of a flexural ultrasonic transducer that has arguably negligible influence on dynamic response), and matching layers in sensor designs. Finally, it should be understood if the chosen transducer materials will be able to withstand the forces or loads associated with the chosen application. This would be especially important to consider for applications including ultrasonic cutting or welding, and can be assessed by revisiting Step 3, before progressing with a revised (but not necessarily completed) model or simulation, towards Step 5.

5. Define a suitable manufacturing process.

Prior to arranging for any component materials to be machined into the required shapes or procuring expensive active driver materials like piezoelectric ceramics for the device assembly, it is important to give due consideration to how the device and wider system (such as support fixtures) will be manufactured. It is highly likely that assistive fabrication tools will need to be designed alongside, and that these will be bespoke to the device. It is a critical but often overlooked step in the manufacturing process of an ultrasonic transducer, but undertaking this properly ensures a high degree of consistency and repeatability in the fabrication process. For example, an alignment tool is generally required for the central positioning of a piezoelectric ceramic disc on the underside of the plate in the case of the flexural ultrasonic transducer. A mechanism is also needed to ensure the piezoelectric ceramic disc is fixed in position with a sufficiently high force to allow the epoxy resin in the transducer assembly to cure to the desired physical condition, in some cases for several hours. These alignment tools usually need to be designed and constructed to enable the fabrication of the transducer, and so proper planning is imperative. One further point of note regarding this is that the production of assistive tools generally assumes the production of only a small number of devices is being undertaken, for example, single transducers or in batches of 10 or 20. This is because fabrication in this way can be time consuming. There are additional challenges for the scale-up of transducer manufacturing, including repeatability across key structural dimensions to ensure consistency in the dynamic performance. It is assumed here that single or small batches of transducers are considered in the interests of the general readership, noting that scaling the manufacturing potential of different classes of ultrasonic transducer beyond currently mass-produced configurations such as the flexural ultrasonic transducer is an important challenge.

6. Undertake a trial transducer fabrication.

Practice makes perfect, and it is highly unlikely that the ideal device at will be fabricated at the first attempt, even if the steps listed here are rigidly adhered to. It is strongly recommended that time and resources are dedicated to trialling the fabrication process, and that the outcomes of every step are recorded and analysed in detail. It is inevitable that by trialling fabrication strategies and implementing revisions and modifications to mathematical models, alternative preferred routes to the manufacture of a high-quality device will emerge for a specific application. It is also why the production of reliable assistive fabrication tools, the details for which are set out in Step 5, is so important. Without these, it will be difficult to properly understand the limitations of the fabrication process being implemented, and it will not be possible to ensure the consistency of transducer fabrication across a batch of devices. This will make the manufacture of matched pairs of transducers, for example, that are vital in sensing applications, difficult. By producing several samples and revisiting the mathematical models where necessary, the optimisation of a fabrication process can be undertaken.

Finally, it is strongly advised that electrical impedance analysis (EIA), or similar, is used throughout the manufacturing process. EIA is an extremely powerful tool for assessing device quality if it is used in the correct way. It can be used to rapidly determine deviations in resonance, monitor electrical impedance, generate admittance loops, monitor other electrical properties such as susceptance and conductance, and generally be employed as a mechanism to benchmark the quality of ultrasonic transducers. It can also save significant amounts of time by revealing faults in the device fabrication process, particularly at each stage.

7. Outline a revised fabrication strategy for implementation.

Using the outcomes of Step 6, a coherent manufacturing approach for the fabrication of the transducer can be defined. Any uncertainties can be addressed by revisiting the earlier steps as outlined above. For example, a common outcome is that an assistive fabrication tool is not practical for ensuring consistency in the fabrication of multiple transducers. The discrepancies in the fundamental resonance frequencies of interest may be undesirably large, requiring modifications to the fabrication approach. By taking the time to think about the results from the trial fabrication as shown in Step 6 to define a suitable fabrication strategy, the variability in terms of dynamic performance from one transducer to another can be minimised, and thereby the potential errors or discrepancies in dynamic performance which often arise in the application phase (and hence usually too late) can be mitigated or eliminated. Using this advice, the manufacture of an optimised iteration of the transducer design can be undertaken, after which dynamic characterisation can be implemented according to the advice provided in Step 8.

8. Conduct a comprehensive dynamic characterisation.

Prior to the trial of an ultrasonic device in its intended environment or application, it is critical to assess all the important operational properties of the transducer and compare with the outcomes of the modelling and simulation approaches as per Step 3. Prior to any experiment, the boundary conditions on the device should be considered, such as how it is fixed in place for testing or for the target application, and how these boundary conditions may influence the dynamic characteristics of the transducer. It is not uncommon to make minor modifications to a model or simulation at this stage, to ensure the prediction of device performance is suitably representative. For example, if the device is designed to be attached to a frame, it should be understood if any part of the device is being constrained in a way that it is not in the mathematical model. A check on the similarity between the mathematical model and the real experimental environment should be made. However, it is not expected that the conditions defined in a mathematical model are identical to those of an experimental setup, but key parameters including environmental fluid, temperature, and pressure in a model or simulation should be set as close as possible to the practical environment.

A recommended first step of the dynamic characterisation process is EIA, and by inspecting key parts of the electrical impedance spectra, it is possible to determine the resonance frequencies of interest for the device, for example, via the series resonances. An optical experimental technique such as laser Doppler vibrometry (LDV) can then be undertaken, to capture the mode shapes or vibration amplitudes of the ultrasonic transducer, where the displacement amplitudes of interest can be obtained for a given excitation condition (this will likely be in terms of a voltage). Other dynamic characterisation approaches used in this step will depend on the device configuration. For example, the measurements of the radiation patterns and acoustic fields for a sensor through an instrument such as a microphone are important for classes like the flexural ultrasonic transducer, designed for proximity sensing and ultrasound wave propagation. From each experimental dynamic characterisation step, correlations can be made with the mathematical modelling outcomes, as per Step 3 and beyond, with any revised iterations implemented. As a further note on the experimental aspects of this step, it should be ensured that all details of the experimental steps are recorded, including the excitation conditions on the transducer used for each technique. The excitation levels administered to a transducer can influence the dynamic characteristics that are measured.

Upon the conclusion of Step 8, it is likely that there will be a relatively high degree of confidence in the transducer and its operational capabilities, providing the modelling, fabrication, and dynamic characterisation have been implemented with the proper controls in place. The outcomes of each step can be used to evidence the suitability of the device to the application for which it is intended. The key principles associated with the steps above will be expanded upon at various points in this chapter, providing guidelines on some recommended practices and indicating several common errors of which the design engineer should be aware. To define the design process schematically for later reference, the cycle of the design and manufacture process for an ultrasonic transducer is illustrated in Figure 3.2.

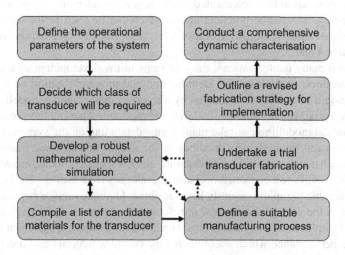

FIGURE 3.2 Summary of ultrasonic transducer design and manufacture, where solid arrows indicate typical progressions between steps, and the dashed arrows indicate where revisions and changes to approach are often required.

Here, the key stages of the design steps are shown, and although there is generally a linear progression through the steps depicted, there are evidently circumstances where the design approach must be revised, with key steps revisited. The requirement for this is highlighted in Figure 3.2, showing particularly common revision stages between key steps designated as the dashed arrows. For example, it is common for observations found during the trial transducer fabrication stage to require a revision of the mathematical modelling approach, often leading to a change in the manufacturing process. Generally, there is a need for strategic planning in the design and manufacture of an ultrasonic transducer because it can be a time consuming and expensive process. In all stages, there should be mitigation procedures defined to resolve simulation or fabrication challenges which are likely to be encountered. An example of a mitigation strategy for Step 6 for instance, may be to fabricate multiple samples in the event of a trial fabrication not meeting the desired standard.

3.3   PRACTICAL ASPECTS OF TRANSDUCER DEVELOPMENT

3.3.1   Defining the System Parameters

The first three steps as illustrated in Figure 3.2 together constitute a scientific or engineering approach to understanding the nature of the system for which the ultrasonic transducer is being designed. The system encompasses the transducer, the environment, and any other structures or components which will come into contact with the transducer or influence its dynamic performance. Therefore, a suitable class of ultrasonic transducer can be selected based on the considerations associated with the target application. This allows a robust mathematical or finite element approach to be implemented. This section considers ultrasonic transducers generally, though the practical advice given can be used in the design and fabrication of adaptive ultrasonic transducers, alongside the information provided in Chapters 4 and 5.

As with many configurations, the first steps in the development of the ultrasonic transducer encompass the identification and proper definition of what the device should be able to do, followed by a rigorous mathematical modelling and simulation process where the key material and mechanical parameters are defined. Mathematical modelling can take many forms depending on the system, but it is common to use finite element methods in the design of ultrasonic transducers. Electromechanical modelling, for example, network models including Mason's model (Mason, 1948), or electromechanical analogues (Dixon et al., 2017), can be highly valuable to rapidly quantify key transducer features, such as required plate diameter in the case of the flexural ultrasonic transducer, or in the derivation of expected amplitude–time relationships. However, for a complete representation of dynamic performance which accounts for the complex material and mechanical properties of a wide range of advanced materials, multiphysics approaches are highly valuable. It should be noted that there are some limitations with finite element methods and mathematical modelling that have affected the progress in the development of adaptive ultrasonic devices, particularly those incorporating shape memory materials. Some of these limitations will be explored in more detail in Chapter 4, but in general these models require accurate material properties such as Young's modulus, and this is not always straightforward to measure for shape memory alloys. Second, it is typical for an ultrasonic transducer design to be ideally defined through finite element methods, for example, optimised with regards to frequency and vibration amplitude. However, the fabricated device then exhibits deviations associated with factors including inconsistencies in the transducer fabrication process that are too difficult or impractical to control. Such discrepancies can typically manifest as deviations in resonance frequency or vibration amplitude output, and dissimilar bandwidths in batches of nominally identical transducers.

To begin, the fundamental purpose of the transducer should be defined. For example, if the transducer is to be used for a power ultrasonics application, then the most likely candidates would be any transducer configuration permitting the application of high-power ultrasonic vibrations to the target system or environment. A typical class of transducer in this case is the Langevin. As shown in Chapter 1, this configuration has been perhaps the most studied form of transducer in the field of ultrasonics over the past century, emerging from its implementation in underwater sound measurement and ranging applications towards dentistry and surgery, to the wide array of medical and industrial applications we know today. As shown in the introductory chapters, the Langevin class of ultrasonic transducer has been applied for ultrasonic welding, orthopaedic surgery, inducing cavitation, and the compaction and consolidation of granular materials for manufacturing, among others. Despite the variety of applications, the fundamental design of this transducer has hardly changed over the past 100 years (Bejarano et al., 2016). Like it did in the early 20th century, the Langevin class of ultrasonic transducer still consists of a tuned front mass, piezoelectric ceramics, and a back mass, in a single assembly pre-stressed using a centrally aligned bolt and which is tailored to ensure sufficiently high vibratory motion of the front mass and end-effector. The transducer is tuned to a specific mode of vibration, and as such there is some freedom in the geometry and material properties of the front mass or end-effector, such that the configuration can be tailored to different applications. Conversely, sensing or lower power approaches can often necessitate the consideration of alternative transducer designs. Early forms of flextensional transducer were designed to radiate sound and operate as sonar-type devices (Rolt, 1990), but more recent developments have transitioned these designs towards power ultrasonics applications, including prototypes for ultrasonic biopsy procedures (Mathieson et al., 2015a), and orthopaedic surgery (Bejarano et al., 2016). An example of the Class V flextensional cymbal transducer adapted into a power ultrasonic device for orthopaedic surgery is shown in Figure 3.3, reprinted from the work of Bejarano et al. (2016).

In terms of the important functions or features of an ultrasonic transducer for a specific target application, there can be several options that are important to consider. For example, flow measurement can be undertaken using a flexural

FIGURE 3.3 An ultrasonic orthopaedic surgical device with a Mectron S.p.A. cutting tip, based on the Class V flextensional cymbal transducer. The cutting tip is attached via a thread to the cymbal end-cap, where the end-cap has a total external diameter of 33.0 mm. Reprinted from the work of Bejarano et al. (2016), under the CC-BY 4.0 licence.

ultrasonic transducer, or alternatively a device fabricated using composite materials matched to the environment. However, the choice between different classes of ultrasonic transducer may depend on restrictions relating to the system or environment. These can include the difficulty of physical access (it is not uncommon for flow measurement through long sections of pipeline to be of interest), or if an environment exhibits harsh physical characteristics including elevated levels of pressure or temperature. In such cases, if one were unable to position a sensor in-situ, then a flexural ultrasonic transducer may not be the optimal choice, where a clamp-on configuration may be preferable. In any case, the list shown in Figure 3.4 outlines what can be considered in the selection or design of a transducer.

The suggested considerations shown in Figure 3.4 can be extended to other transducer classes, and there may be additional factors to consider as particular to a configuration of ultrasonic transducer. The aim of Figure 3.4 is to capture the important considerations for most transducer designs. Once these questions are addressed, the selection of suitable materials and a viable mathematical or finite element modelling approach can be implemented, followed by the implementation of a transducer fabrication process.

3.3.2 MATERIAL SELECTION

The choice of materials for the selected ultrasonic transducer configuration is dependent on both the intended application and the transducer class. It is necessary to consider the manufacturing process in each case, factoring in how straightforward it would be to machine a transducer component into the shape suitable for incorporation in the transducer design. For example, the shape memory alloy Nitinol can be extremely difficult, time consuming, and expensive to accurately machine into complex shapes, and certainly those more intricate or complicated than cylinders or sheets. These factors are explored in more detail

Consideration	Questions to Ask
Boundary Conditions	What are the boundary conditions arising from the application?
Dynamics	Do I require narrowband or broadband?
	Which vibration modes do I need to exploit?
	Will modal coupling be required?
Environment	Will the transducer be operated in situ?
	If not, how will it be integrated with the system?
	If so, what are the temperature and pressure limitations?
	Are there other environmental or physical implications?
Frequency	What are the required operating frequencies?
Power	Is the transducer for a power ultrasonic application or for sensing?
Production	Do I need to make a few transducers, or many?
Size	There is often a close interdependency of size and dynamic properties such as resonance frequency.

FIGURE 3.4   A sample of key considerations for transducer selection.

later in this book. A more general list of considerations for the materials can be regarded as the following.

1. **How will the transducer be driven?**
   i. If the transducer design is to incorporate a piezoelectric element as the principal driving mechanism, then one must consider the operating temperature range of the application, the amplitudes required (directly obtainable from the piezoelectric properties), and whether hard or soft materials should be used. Generally, harder piezoelectric ceramics are required for power ultrasonics applications, and softer varieties are needed for low power sensing. There are a wide range of piezoelectric ceramics available from manufacturers, and so it is strongly recommended that requirements are discussed with them during the transducer design process.

ii. If the transducer is to be driven via an alternative mechanism, for example, electromagnetically such as in the case of EMATs, then the electromagnetic coil configuration should be clearly defined. There are often only a few material options available in this case, which should be thoroughly reviewed.

iii. Alternative driving mechanisms are available, including the use of magnetostrictive materials. Similar considerations are applicable as item i above in this list, but there can be different demands on the excitation system. For example, these can include higher current levels compared to the voltage demands of a piezoelectric-based system.

2. How will the ultrasonic waves propagate through the device?

i. Acoustic and ultrasonic waves will attenuate depending on the properties and features of the materials they are travelling through. Some materials will absorb in the form of heating, and others will scatter ultrasound waves. This can also occur by other means, for example, via small flaws, cracks, or impurities inside a component.

ii. It is often desirable to design a transducer to operate with multiple vibration modes coupled together torsional (Al-Budairi et al., 2011, 2013; Cleary et al., 2022). This can be undertaken to deliver specific motions, for example, combining longitudinal and torsional modes in the Langevin transducer for optimal cutting performance.

3. How will the ultrasonic transducer be fabricated?

i. Like the comments in the introductory material above, it is important to strategise the manufacture of any part or feature of an ultrasonic transducer that is obstructively difficult to machine or define. Occasionally, the approach is not practical to justify and so an alternative fabrication process is often implemented.

ii. If any hot working is required to machine a component, it is important to gain an understanding of how the process may induce permanent changes to the material properties of the component, thus influencing the ultrasound wave propagation through the transducer. The dynamic performance of the ultrasonic transducer in general would typically be affected, and so all aspects of the fabrication process should be assessed in this regard.

4. What amplitude and stress levels will the transducer likely experience?

i. Fatigue is a notable concern for the longevity of ultrasonic transducers, particularly those operating in power ultrasonics applications, for example, due to the relatively high levels of cyclic loading to which certain components are exposed. Materials with higher fatigue lives or limits are desirable in such applications, such as titanium and aluminium alloys in the case of power ultrasonic transducers. There is a degree of flexibility in the selection of materials for certain sensor configurations. For example, a wide range of metallic and non-metallic materials have been incorporated into the flexural ultrasonic transducer.

ii. Even if the specific amplitude or stress limits are not known at the outset, finite element methods can be extremely useful to understand safe operating estimates and to engineer a transducer design with an acceptable factor of safety.

5. **Are there specific environmental factors to consider?**
    i. For example, if the transducer, or any of the constituent components, will be exposed to elevated temperatures, then suitable materials should be selected to withstand those conditions.
    ii. In certain applications, such as flow measurement or gas monitoring, phenomena such as corrosion must be considered, and therefore materials with passive layers including titanium alloys are regarded as particularly suitable, possessing physical characteristics, such as density and Young's modulus, at magnitudes desirable for operation at low ultrasonic frequencies (commonly below 100 kHz).
    iii. There are various regulations to consider, and some materials contain hazardous elements, such as lead in the case of lead zirconate titanate piezoelectric ceramics. Materials are often subject to controls, and this can be linked to the environment.

Given that many of the transducer classes discussed in this book typically incorporate piezoelectric ceramics, it is therefore necessary to include a brief overview of piezoelectric materials with some general advice on the ways in which they can be incorporated into an ultrasonic transducer. This material is provided in the next section, alongside some physical properties to be aware of for the design and fabrication process.

3.3.3 PIEZOELECTRIC MATERIALS

For the purposes of the concepts discussed in this section, it is assumed that the reader is familiar with piezoelectric materials and the fundamentals of their operation. In general, there are many compositions of piezoelectric material now available from leading manufacturers, encompassing an array of forms suited to different configurations of transducer. Therefore, it can be difficult to determine an optimal composition of piezoelectric material for an ultrasonic transducer design. This section aims to outline some of the key factors to consider, though the principal lesson to take away from this section is that consultation with experts in the field is always strongly recommended. It is also noted that many readers will likely have access to advanced fabrication facilities, well equipped with the expertise required to synthesise the necessary compositions of material and fabricate piezoelectric elements to the necessary quality. Typically, piezoelectric elements can be commercially procured in a wide range of bulk forms, commonly as discs, rings, plates, and tubes, with poling along specific designated coordinates depending on the geometry and the intended application. A selection of common configurations (with examples of typical target ultrasonic transducer classes) include the following:

- Discs, for ultrasonic transducers such as PMUTs and flextensional types.
- Rings, for Langevin transducers.
- Spherical or hemispherical, for sonar, hydrophones, and HIFU.
- Plates, for flextensional transducers.
- Tubes, for power ultrasonic applications including those requiring cavitation.

More advanced forms of piezoelectric element are also available, such as those embedded with polymers to operate as flexible patches (Zhao et al., 2021). These materials are particularly well suited to some medical applications such as wearable technology. The configurations in the list above can also be combined to construct novel actuators for a variety of applications, for example, those in robotics.

Once a suitable transducer class is identified for a given application, a candidate configuration of piezoelectric material can then be selected, using the list above as a reference. It should be noted that some applications require specific geometrical considerations, and it is often necessary to machine or tailor the piezoelectric material to an appropriate size, if one cannot be procured through commercial means. Bespoke piezoelectric materials can be synthesised to customise the material properties through the material composition, and the operating mode and dynamics via the physical size and shape of the element. It is often necessary to dice and lap bespoke elements to suit the device into which the piezoelectric element will be embedded.

The next step to consider is the composition of the piezoelectric material itself. Many power ultrasonic applications utilise bulk piezoelectric elements, whereas many configurations of ultrasonic sensor, though not all, often incorporate composites which are designed to acoustically match to the external environment or other transducer components, specifically with regards to the acoustic impedance property. Irrespective of which approach is followed, the composition of a piezoelectric element is critical to its suitability for a given application, where the chemical elements within a piezoelectric composition directly influence the operating performance of the material. For example, there continues to be much discussion surrounding the inclusion of the chemical element lead in the popular lead zirconate titanate (PZT) piezoelectric ceramics, principally regarding its toxicity to humans and the environment (Bell and Deubzer, 2018). Despite this, it remains a popular choice for transducer designers, in part due to its achievable output, particularly with regards to the d parameter, the polarisation property which is directly proportional to unit stress or strain. The methodical selection of a suitable piezoelectric material composition is important, because even within lead-based variants there is a wide array of options available. The choice of piezoelectric material can be made simpler by comparing different materials across several key properties. It is advisable to consider the following at the outset:

1. **The frequency constants**, which are often designated by N and are the product of the geometrical dimension associated with the vibration mode of interest, and the resonant frequency.
2. **The Curie temperature**, T_C, above which the piezoelectric (and ferro-electric) properties of the material are lost. It should be noted that in practical terms, the viable maximum operating temperature of an ultrasonic transducer should not be aligned directly with the Curie temperature of the piezoelectric element embedded within it. This can be around half in practice, though this is a guideline and can be empirically determined.
3. **The dielectric loss**, which is often designated as *tanδ*. This quantifies the energy loss in the material because of heating, for example, from the electric field.
4. **The piezoelectric charge constants**, designated by d.
5. **Mechanical quality factor**, Q_M, to provide an accurate indication of damping.
6. **Electromechanical coupling factor**, k_{eff}, which is the effectiveness by which electrical energy can be converted into strain in the material.

Prior to transducer manufacture and in the early stages of the design process, it is often customary to conduct a thorough review of available piezoelectric materials, configurations of element available, and how readily these elements can be integrated into the proposed transducer configuration. Dependent on the composition of the piezoelectric elements in a transducer design, bulk piezoelectric materials can be broadly split according to their hard or soft types. Harder types of piezoelectric ceramic are those more suited to higher amplitudes of vibration, such as those associated with power ultrasonic transducers or cavitation generation. Softer piezoelectric materials, achieved through approaches such as introducing higher proportions of niobium in the PZT composition (Boota et al., 2023), are fabricated for their enhanced sensitivities, commonly used as sensors. Harder piezoelectric materials typically consist of higher quantities of elements such as iron and chromium (Boota et al., 2023; Ketsuwan et al., 2009), and they can exhibit significantly higher quality factors than their softer counterparts. Therefore, in the design of an ultrasonic transducer to be used as a sensor, a softer variant of a PZT would generally be more suitable for the design. In general, manufacturers of piezoelectric materials commonly provide the most useful advice on the advantages and disadvantages of the options available, but this usually requires consultation because applications can have specific requirements that only certain piezoelectric materials will be able to meet. It is not always possible to do this by reading a specification sheet alone. Most manufacturers of piezoelectric materials will tailor their products towards different applications, including but not limited to, accelerometers, flow measurement, medical imaging, NDT, pressure monitoring, power ultrasonics, cavitation generation, and sonar. There are specific requirements associated with each of these applications, and so a detailed consultation with manufacturers is always advised.

One other important aspect of piezoelectric elements to consider for transducer design concerns the electrodes which facilitate the application of an excitation signal to the element. These electrodes which are typically deposited on the surface of a piezoelectric element are often screen printed to the elements using materials such as silver. This allows electrical connection across the coordinate of poling, and the specific patterns of electrodes which are commercially available are often varied. For example, it is possible to procure piezoelectric elements with electrodes deposited over a whole surface or face of the element, or they can be screen printed on specific sections only. Alternatively, it is also possible to screen print in specific and more complicated patterns, including wrapping an electrode around an element. A pattern like this is particularly well suited to piezoelectric elements for integration with the flexural ultrasonic transducer, as it permits the application of an electric charge across the element, from one point of access. This means that the element can be bonded to a plate or membrane without the need to engineer an intricate or complex mechanism to apply the electric charge via the membrane or plate structure.

In terms of piezoelectric materials, it can now be assumed that there is an understanding of the selection process for a suitable portion, or series of elements, for the ultrasonic transducer to be designed. Given this information and how the understanding of both the system and the materials has been defined, it is henceforth necessary to provide an overview of how mathematical modelling and finite element analysis can be employed in the transducer design process.

3.3.4 Mathematical Modelling and Simulation

The mathematical modelling or finite element analysis of an ultrasonic transducer is a fundamental component of transducer design. There are many mathematical modelling processes and tools available, and some thought should be given to the choice of modelling strategy which is best suited to the application. Some options involve analogue representations of transducers to rapidly generate an equivalence for the device from which key dynamic features can be identified. Conversely, some approaches utilise a broad range of material properties and boundary conditions which are defined with precision, in models from which key multiphysics phenomena can be extracted. This section briefly outlines two approaches which have become popular for practical ultrasonic transducer design and manufacture. The first is the *equivalent circuit technique*, and the second is *finite element analysis*. These approaches yield different sets of data and are hence not two options to obtain the same outcomes. Their capabilities should be understood prior to their application.

3.3.4.1 Equivalent Circuits

The application of equivalent circuits in developing models of the responses of ultrasonic transducers is based on the rationale that mechanical characteristics of a system possess electrical equivalents. For example, the velocity parameter in a mechanical system can be reasonably understood to correlate with electrical current. Similarly, the force on a mechanical system can be considered as analogous to an electrical voltage, and mass can be considered as equivalent to inductance in yet another similar way (Stansfield and Elliott, 2017). Given that ultrasonic transducers are effectively electromechanical devices, it thus follows that if an ultrasonic transducer is embedded as part of a system in an application, it is possible to consider the ultrasonic transducer as an electrical load. From here, it is then possible to construct a representative equivalent electrical circuit for an ultrasonic transducer. One key purpose for doing this is that given the analogies between mechanical and electrical quantities, further comparative quantities can be considered. Since parameters such as velocity and force have electrical equivalents, mechanical impedance possesses an electrical equivalent, in terms of electrical impedance. Therefore, a simplified representation of mechanical acoustic impedance can be obtained through modelling the ultrasonic transducer as an electrical circuit, thus determining the electrical impedance. These quantities are invaluable for understanding the dynamic and resonance characteristics of ultrasonic transducers, and equivalent circuit modelling approaches have presented significant opportunities for understanding the underlying mechanisms of complex dynamic systems. Fundamentally, and with relevance to piezoelectric-based ultrasonic transducers, it will be necessary for any equivalent circuit to incorporate a representation of the piezoelectric material, the excitation (or input voltage and current) to the ultrasonic transducer, and the output characteristics.

In the engineering of ultrasonic transducers, Mason's model is commonly used for systems incorporating piezoelectric materials, in part because it tends to yield reliable information regarding the electrical–mechanical coupling. An example of the equivalent circuit for a piezoelectric ultrasonic transducer is shown in Figure 3.5, extracted from the work of Afzal et al. (2018), showing the direct translation between the transducer assembly and the electrical equivalent.

The multilayer transducer shown is for an ultrasonic sensing application. The equivalent circuit as depicted in Figure 3.5(b) considers the piezoelectric properties of the PZT layer shown in Figure 3.5(a), in addition to the boundary conditions on the transducer, in this case from the pressure from the external environment on the radiating transducer surface (Afzal et al., 2018). The capacitance and inductance in the equivalent circuit can clearly be observed via the standard symbols in Figure 3.5(b). It should be noted that this equivalent circuit model, whilst used as an illustrative example here, neglects acoustic or electrical damping in the interests of simplicity.

Equivalent circuits of this nature can be used to generate a range of outputs, such as impedance–frequency spectra or admittance loops. Admittance is the inverse of impedance, and it is a measure of how easily electrical current might flow in a system. Just like impedance, it contains both real and imaginary parts which can be experimentally measured. These are conductance and susceptance. The measurement of these properties, for example, with an impedance gain/

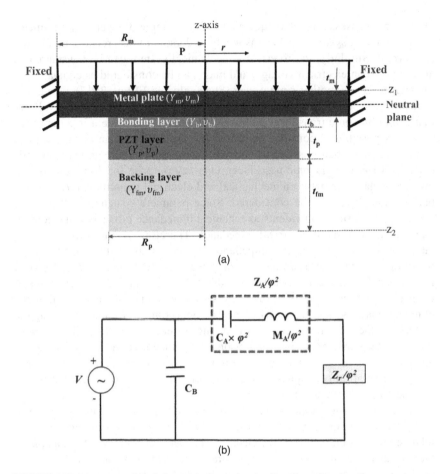

FIGURE 3.5 An example of the equivalent circuit of a piezoelectric ultrasonic transducer, showing (a) a multilayered transducer concept, and (b) its electromechanical equivalent circuit, where Z_A is the acoustic impedance, Z_r is the radiation impedance, V is the voltage, M_A is the acoustic mass, C_A is the acoustic compliance, C_B is the electrical capacitance, and φ is the turning ratio. Reprinted from the work of Afzal et al. (2018), under the CC-BY 4.0 licence.

phase analyser, can then be used to generate admittance loops, as shown, for example, in the scientific literature for the flexural ultrasonic transducer (Feeney et al., 2017a). These loops can be used to extract useful information about the ultrasonic transducer, including the series (resonance) and parallel (antiresonance) frequencies. The equivalent electrical circuit and the corresponding admittance loop are hence closely connected. Depending on the configuration of ultrasonic transducer selected for the target application, modifications to the equivalent electrical circuit can be made, and corresponding influences on the admittance loop monitored.

3.3.4.2   Finite Element Analysis

The ability to predict, with a reasonably high degree of accuracy, a variety of physical properties and behaviours of an electromechanical system incorporating an ultrasonic transducer is extremely valuable, in terms of both practicality and cost. The ability to capture the multiphysics phenomena of an ultrasonic transducer in advance of fabricating it has been revolutionary in the past few decades. In general, it allows resonance frequencies, modal behaviours, displacement amplitudes, stresses, and other properties including important electrical parameters to be obtained for a given excitation condition and set of boundary conditions. Modern finite element analysis software can rapidly solve highly complex problems, and many developers now offer cloud-based solvers which remove the need for costly and intricate server systems which formerly needed to be locally based. It is strongly advised that a robust finite element model is produced prior to any fabrication taking place. A high-quality model can be very informative for the transducer design process, and which can be relied upon as a model which can be taken forward and applied to a variety of conditions or applications. In this section, the finite element approach is briefly introduced in the context of the design of piezoelectric ultrasonic transducers, with advice and recommendations which are important to consider for the practical phase of transducer assembly.

In general terms, it should be acknowledged that the value or reliability of a finite element simulation is dependent on the quality of the inputs provided. It is strongly advised that accurate material properties for each associated component in the transducer are acquired. As a minimum, the fundamental material properties one would require for performing simulations of modal behaviour include, but are not limited to, Young's modulus, mass density, Poisson's ratio, and material properties specific to piezoelectric materials including the charge and voltage coefficients. Material properties can be experimentally obtained, for example, by making use of universal testing machines, or sourced from manufacturer data. Material properties can vary in accuracy from one sample to another, and it is often good practice to obtain the necessary data experimentally if possible. The principal challenge for defining material properties in a finite element model, irrespective of the software or code employed, is that there is a dependence on the manufacturer of piezoelectric materials to provide accurate and reliable data for their products. For several reasons, this data may not be available, and so there are options available to address this. One is to measure the properties of the piezoelectric materials using a technique such as EIA. The second is to consider the class of piezoelectric material for use in the device, and to approximate it with the properties of similar products that are available. For example, there are many variants of soft lead zirconate titanate piezoelectric ceramic, one of which is PZT-5H. This piezoelectric ceramic material is often used in lower power sensing devices, such as the flexural ultrasonic transducer. Occasionally, the precise properties of a material such as this may be required, but the data at the sufficient level of accuracy is not available from the manufacturer. In such cases, there can be workarounds, for example, a PZT-5H piezoelectric ceramic is a Navy Type VI material.

This means that other types of piezoelectric ceramic also classed as a soft Navy Type VI may be used as an approximation in a representative finite element model. This may be an expedient way of producing a finite element model within generally acceptable margins of error. This can also apply if a transducer designer is working with piezoelectric materials synthesised with bespoke methods, for example, in a laboratory. Clearly, great care must be taken with this because the properties used for the simulation may not be wholly accurate. Nevertheless, the Navy Type classifications are highly valuable for design and modelling of transducers, though it is advised that they are reviewed in the transducer design process.

It is also important to be aware that materials, and the transducer components manufactured from them, are not inherently ideal in their composition or mechanical structure. They typically all contain impurities, imperfections, and asymmetries to differing degrees, irrespective of whether there are robust manufacturing controls in place, which can all impart notable influences on the dynamics of a transducer fabricated from them. This is a further reason for characterising the properties of materials prior to integration in an ultrasonic transducer, the inclusion of the information in an associated finite element model, and for undertaking an assessment of the structural and physical condition of each component that is selected for a transducer.

Importantly, and specific to the finite element modelling process, one should carefully consider the purpose of finite element modelling for the application of interest. Fundamentally, finite element simulation enables a highly accurate prediction of the behaviour of the ultrasonic transducer, in terms of mechanical characteristics such as stress and displacement, and electrical properties including impedance and (output) voltage. However, finite element analysis can also be utilised as a valuable tool to aid in the physical manufacture of the transducer, where it could be astutely used to align physical measurements made in the transducer manufacturing process with those of what the simulation of the transducer predicts. For example, if the influence of an applied pre-loading condition on the dynamics of a Langevin transducer during manufacture requires assessment, the pre-loading condition could be defined in a suitable finite element model, after which the analysis for a particular output parameter could be implemented. This could be electrical impedance on the piezoelectric ceramic stack, for example. In this case, an experimental process could be used to gather measured values from an electrical impedance analyser. This can therefore provide a useful and rapid way of validating a finite element model, and to verify the robustness of the transducer manufacturing process. However, to reach this point, a series of considerations are important:

1. Are the material properties accurate?
2. Which code-specific element types are needed? These are particular to the software of choice.
3. Which simulation routines are of interest? Usually, for the designers of ultrasonic transducers, these are predominantly a frequency analysis (for identifying modal frequencies and mode shapes) and a modal analysis

(often conducted at steady state, where this analysis routine can be used to generate absolute data including displacement amplitudes, for a given excitation voltage condition to the piezoelectric ceramic stack).

4. What are the boundary conditions on the transducer, relevant to the chosen simulation routine? Some routines will not permit certain boundary conditions to be applied, depending on the software used, and so care should be taken in this process.

5. Leading from the points above, which operating modes are of interest? For example, it is common for the modal response of Langevin transducers to be tuned to the first longitudinal mode of vibration, often referred to as the L1 mode. However, there is growing interest in the coupling of modes together, such as longitudinal and torsional (Al-Budairi et al., 2011, 2013; Cleary et al., 2022). With continued innovation in adaptive ultrasonics, research into this area will likely progress in the next few years. Regarding modal frequencies, it should also be considered that it is usual for a sufficient frequency separation between resonant modes be engineered. This is critical for power ultrasonics applications where elevated levels of excitation voltage are utilised, where the risk of leaking energy into other modes, or modal coupling, substantially grows as the excitation voltage levels increase. This is primarily due to nonlinear dynamic behaviours, and this is discussed in greater detail in Section 3.3.6.

6. As an extension to point 4, is it important to conduct a comprehensive mesh convergence analysis of the finite element model? In most cases, this should be undertaken. A high-quality finite element model should allow sufficiently high resolution of data across the fundamental modes of interest, but also higher order modes. To realise this, a mesh optimisation strategy can be implemented. A finite element model with too few elements will not yield acceptably accurate results. However, an excessive quantity of elements in a finite element model can cause the solve time for the simulation to significantly increase. Critically, this can occur for no notable improvement in the accuracy of the simulated data, and so the balance between element number, or mesh density, and the accuracy of the simulated data. It is therefore common to implement a finite element simulation to assess the calculated frequency for a given mode, as a function of the number of elements in the model, or the mesh density. This can be implemented for multiple modes of vibration. As shown in the example in Figure 3.6 (Feeney, 2014), the modal frequency will eventually approach a steady magnitude, from where an optimised mesh density can be defined which satisfies multiple modes of interest. Here, the mesh density, or number of elements, can be increased until the resonance frequency for a particular mode of vibration approaches a steady magnitude. This ensures that the mesh density of the finite element model is optimal in terms of computation time and accuracy. This mesh convergence process should be conducted on several vibration modes of interest.

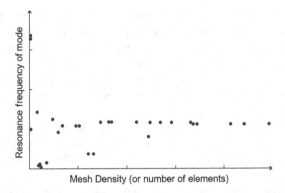

FIGURE 3.6   Example data showing mesh convergence for a generic ultrasonic transducer, using data from Feeney (2014).

Through these considerations, it is possible to generate a high-quality finite element simulation for the piezoelectric transducer configuration of choice. As a minimum, frequency and modal analysis simulations should be undertaken to capture
the relevant data for the experimental dynamic characterisation of the transducer.
Furthermore, a reliable finite element model is invaluable for progressing towards
more advanced configurations of transducer. From this stage, the manufacturing
process of the transducer can be defined.

3.3.5   Tailoring the Manufacturing Process

Up to this point, the system, the boundary conditions, the transducer configuration
most suitable for the application, the materials and components, and the mathematical modelling process should all have been considered, with any possible preliminary
prototyping conducted as required. And as would be expected, the manufacturing
approach which should be selected primarily depends on the configuration of ultrasonic transducer and the constituent materials in the transducer assembly.

It would not be practical to disseminate the typical manufacturing approach for
every configuration of ultrasonic transducer referred to in this book, but a few
examples are included in this section as a guide to how one might begin. As a note,
the extensive bibliography provided near the end of this book contains high-
quality publications on all aspects of ultrasonic transducer designs across the
classes detailed in Chapter 1. The reader is invited to review this list and consult
the relevant publications of interest. Returning to the manufacturing process and
the ways in which it can be tailored, the first stage is to consider what the fabrication process should look like, including how any elements of this process will
impact on the mechanical integrity of the individual transducer components.
Where possible, manufacturing should consider scale-up, including and how
larger quantities of devices can be produced, not just in singular form or in small
batches. Therefore, ease of manufacture, repeatability, operational lifetime, and
unit costs are important factors.

If the general manufacturing process for a Langevin transducer is considered here as an example, the principal consideration is the pre-stressing of the piezo-electric ceramic stack. This is a critical step in the fabrication of Langevin trans-ducer, which is typically undertaken to reduce the energy loss in the transducer and ensure optimal resilience to higher amplitudes of vibration (DeAngelis et al., 2015; Mathieson and DeAngelis, 2015; Li et al., 2019). This is because piezoelec-tric ceramics such as lead zirconate titanate are relatively poor in tension, thus if they are compressed, they can be driven under higher levels of stress than they would be if they were not pre-stressed. Given the pre-stressing procedure is cen-tral to the manufacture of a Langevin transducer, it is customary to fabricate sup-porting mechanical components to hold the front and back masses in place, whilst the required torque is applied to the central bolt. This can be calculated using Equation (3.3) as extracted from Ranjan et al. (2013), where a typical pre-stressed Langevin transducer is illustrated in Figure 3.7, showing the notches machined into the end-masses to allow the application of torque to the central bolt.

$$\tau = KFd \tag{3.3}$$

In Equation (3.3), the required pre-stressing torque level τ can be estimated through the product of the (pitch) diameter of the bolt d, the applied force F, and the relevant engineering constant K which is generally linked to the material type and is a frictional-related torque coefficient. For example, a zinc-coated steel bolt may have a K constant of 0.2, or in that region. If the estimation of force F is required, then an assumed torque level can be made, or the process undertaken empirically. Then, the applied stress σ from the pre-stressing process on a given target area, such as a piezoelectric ceramic, can be calculated through the ratio of force to the area on which the stress is applied, via Equation (3.4).

$$\sigma = \frac{F}{A} \tag{3.4}$$

Such required torque levels can typically be determined from experience, but it does also depend on the component materials in the transducer assembly.

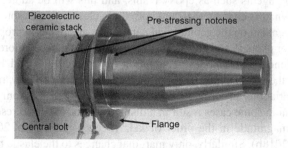

FIGURE 3.7   A pre-stressed Langevin transducer, showing the notches machined into the end-masses to facilitate the application of a pre-stressing torque via the central bolt.

One useful approach to follow in the pre-stressing of a Langevin transducer is to monitor the electrical impedance response of the transducer throughout. So long as the finite element simulation is repeatable and correlates well with the measured data from prototyping, it is possible to acquire a reasonably strong idea of what the resonance frequency of the first longitudinal mode should be. Knowing this, increasing levels of torque can be applied to the transducer assembly through the central bolt, and it will be possible to monitor the progressive change to the series resonance frequency of the transducer at that modal frequency, and the associated electrical impedance. Once the electrical impedance approaches a steady magnitude, the transducer can be considered as suitably pre-stressed. However, an important disclaimer is that some follow-up dynamic characterisation, outlined in the next section, is strongly advised to assess the stability, and verify the expected dynamic performance of the transducer. Through experience, it is known that Langevin transducers, and other configurations especially those used for power ultrasonic applications, benefit from a *settling* period, where the physical properties of the piezoelectric stack and the end-masses equilibrate after the pre-stressing procedure. For example, there is evidence that there can be some mechanical relaxation phenomena in the transducer (Liu et al., 2023a), after which time the dynamic response, including the resonance frequencies and electrical impedances, settle. A settling period can be tailored to the transducer and empirically determined, but a viable option is to drive the transducer at resonance for a set number of cycles over several days and monitor the electrical response of the transducer after each set of cycles. Changes in series resonance frequency and associated electrical impedance can then be recorded, where again there should be an approach to a stable set of parameters. If we return to earlier comments regarding manufacture, and the considerations required to produce either a single device or many (e.g., greater than 10), it is assumed that the reader will be focused on the design and fabrication of one transducer. It is evident to see how commercial production costs could be high for the manufacture of large quantities of Langevin transducers, given the complex steps involved in their assembly requiring high levels of precision, such as the pre-stressing of the piezoelectric ceramic stack.

Regarding flextensional and flexural type ultrasonic transducers, there are different considerations, primarily because conventional configurations involve the use of bonding agents such as epoxy resins, and this will be addressed later. In general, the dynamic characteristics of both these classes of ultrasonic transducer are highly dependent on the geometrical parameters and material properties of the vibrating cap or plate. Given these transducer classes tend to be tailored for operation in their fundamental modes of vibration between 20 kHz and 100 kHz, the tolerances on the physical properties of the caps are extremely low, meaning that only slight variations in terms of geometry, often less than 1 mm for a given dimension, can in some cases result in a disparity in the expected resonance with the nominal magnitude in the order of kHz (Feeney and Lucas, 2018; Feeney et al., 2018a, 2018b). Similarly, only marginal changes to the elastic properties of the cap, such as Young's modulus of the material, can cause a notable shift in the resonance frequencies and achievable amplitude responses. For these reasons, highly accurate manufacturing of the caps is required, and is one reason why contemporary metallic additive manufacturing approaches are desirable. Such methods can potentially deliver components with accuracies beyond what is possible using conventional machining methods, given the dimensions of the caps. However, the required research has not yet been undertaken or significantly implemented for these classes of transducer, in order to assess the dynamic properties of sensors using these additive manufacturing approaches. However, there are important studies available which provide accounts of the characteristics of Langevin transducers fabricated using additive manufacturing (Mathieson et al., 2019), showing that there is strong potential for their use for both medical and industrial power ultrasonic applications in future. One key development in recent years has been the demonstration of the suitable fatigue life of Langevin transducer end-masses for viable operation in typical applications of interest. It should be noted that despite such developments, this research continues to progress.

There can be significant manufacturing challenges for flextensional and flexural ultrasonic transducers with regards to the required tolerances for the caps, plates, and membranes, since even sub-mm deviations in design parameters (such as the diameter or thickness of a flexural ultrasonic transducer's vibrating plate) can lead to kHz scale deviations in resonance frequency. The second major consideration for these transducers is the adhesive bonding. It is typical to bond a piezoelectric ceramic disc, or an equivalent driving mechanism, to the caps in each case. However, the deposition and cured condition of the epoxy resin has been shown to significantly influence the dynamic characteristics of the transducer (Feeney et al., 2019a). The selection of a suitable bonding agent is the first main consideration. For example, a decision should be made if the agent must be insulating or conductive. If the latter, then silver-loaded epoxy resins are widely available and relatively inexpensive, though they tend to exhibit a lower bonding strength than many insulating alternatives.

The next step is to consider if thermal curing is required, and if so, it should be clarified if the properties or mechanical characteristics of the other components in the transducer allow for this. For example, a thermal curing process involving temperatures which exceed the Curie temperature of the piezoelectric ceramic in the transducer should be avoided. It is likely that an alternative piezoelectric ceramic material should be sought, or more practically, a different bonding agent. Next, the fabrication process should be set out in clear steps, including how the components can be integrated into the transducer in a logical and sequential manner. If a bonding agent is used, it is typical for pressure to be applied to the bonding area during a cure. This should be performed in a way that does not lead to deformation or damage to the piezoelectric ceramic disc or the membrane, plate, or cap of the transducer. One way to achieve this is to fabricate a bespoke rig which will accommodate the transducer assembly and allow for a uniform and stable pressure to be applied to the transducer at the required locations. It should also be considered if a vacuum process is required to remove air bubbles in a bonding agent mixture prior to deposition on a transducer component for curing. Air bubbles commonly accumulate in epoxy resins and other bonding agents, and this can compromise the operational amplitudes which are achievable for a transducer over time, and the general stability in terms of resonance and amplitude response.

With these considerations regarding bonding agents, the general advice is to carefully read the agent manufacturer's guidance and integrate it into the transducer fabrication process. Writing out the fabrication process as a logical step-by-step guide in advance of trialling it for the first time is strongly advised. It should also be noted that there are many reports of the challenges associated with bonding agents in ultrasonic transducers, particularly with regards to the dynamic stability of the transducers, their useful life, and the physical integrity of the bond layers. There is ongoing research into alternative manufacturing methods for some of these transducer classes, including direct deposition of active materials on to transducer membranes or plates, thus bypassing the need for bonding agents, for example, using zinc oxide (ZnO) as the active layer (Li et al., 2017). However, there can be costs to achievable amplitude responses, and so each step should be carefully considered.

3.3.6 Dynamic Characterisation

The dynamic characterisation of an ultrasonic transducer, whether it is tailored to a power ultrasonics application at low ultrasonic frequencies or a nondestructive sensing application, can be undertaken by following a general process which is designed to assist in the identification of key transducer parameters, but also in the rapid identification of manufacturing concerns. In this section, an overview of the dynamic characterisation process is demonstrated, with advisory information in support of future developments in ultrasonics, including adaptive ultrasonic transducers incorporating advanced materials such as shape memory alloys and metamaterials.

First, several laboratory instruments are required to undertake the dynamic characterisation of an ultrasonic transducer. For this, the parameters which must be measured should be decided, since each will provide the engineer with different categories of information relevant to the system. Key instruments of interest to engineers of ultrasonic transducers are outlined here, with a selection of important capabilities detailed.

3.3.6.1 Electrical Impedance Analysis

An electrical impedance gain/phase analyzer is a fundamental instrument of importance to ultrasonics scientists and engineers. The high-end instruments are typically expensive, into the thousands of GBP, but should an engineer have access to one then it can make the entire transducer fabrication and characterisation process far more straightforward. There are equivalents available, with fewer features, which are nevertheless useful which can also be utilised. Fundamentally, an electrical impedance analyzer will allow the measurement of several electrical properties of a device to be captured, including but not limited to, conductance, resistance, susceptance, impedance, and phase, all as functions of frequency or other selected parameters.

Primarily, EIA can be used to measure the resonance frequencies of a transducer, via the series resonance locations in an electrical impedance spectrum, but it can be used in a much more informed and useful way throughout the manufacturing process, in part referred to in Section 3.3.4. For example, if used as part of the assembly process of an ultrasonic transducer, the quality of the device fabrication method can be assessed. This will be treated in more detail in the case study at the end of this chapter. EIA will not give much in the way of useful information for the modal behaviour of a transducer, but it can provide valuable information on resonance and where in the frequency spectrum these appear. If one were to fabricate a cymbal transducer, which consists of two end-caps enclosing a piezoelectric ceramic element, an engineer could use EIA to monitor the quality of the epoxy resin bond layers constituting the mechanical coupling in the device throughout the assembly process. The emergence of a splitting of where the expected fundamental resonance frequencies are, informed via finite element analysis, would be detectable through EIA and would form a method of quality control. Using EIA can yield useful information about an ultrasonic transducer, given the required information. Through inspection of series f_s and parallel f_p resonance frequencies, the effective coupling coefficient can be generated, for example, through Equation (3.5).

$$k_{eff}^2 = \frac{f_p^2 - f_s^2}{f_p^2} \qquad (3.5)$$

This relationship, which is strictly related to piezoelectric materials, indicates how effectively the electrical energy in the system is converted into strain, and its associated inverse. From here, other properties or parameters become relevant, including the mechanical quality factor Q_m which can be indicated through Equation (3.6).

$$Q_m = \frac{1}{\tan \delta} \qquad (3.6)$$

Here, the $\tan \delta$ parameter is called the mechanical loss factor, and it is a measure of the damping in the system, where it is particularly useful for when the system is undergoing dynamic loading. In straightforward terms, it is a measure of the stored energy in a material or system as a ratio with the total energy in the system, provided per a unit or cycle. Therefore, a high Q_m system would be one with a high level of efficiency and resonant dynamic response.

It has already been established in this book that the mechanical amplitude response of a piezoelectric ultrasonic transducer has a dependency on the excitation voltage provided to the device. The implication of this is that care must be taken to ensure that the data captured through EIA is carefully and accurately recorded. Comparing measurements from EIA, for example, the resonance frequency through identification of a local minimum in electrical impedance, with resonance frequency determined using an optical method such as LDV, should generally be avoided. First, it is often the case that the boundary conditions applied to the subject transducer are different for one experimental characterisation approach compared to another. Second, and in direct relation to the statement above regarding amplitude response, the excitation conditions for certain experimental approaches can differ substantially to others. It is usual for EIA to be conducted at very low excitation voltage levels, in some cases under 1 $V_{p\text{-}p}$. Conversely, harmonic analyses can be performed in the tens to hundreds of volts, depending on the circumstances. These differences in voltage level have a direct impact on the nature of the amplitude responses of a piezoelectric transducer, as functions of frequency. It is therefore advisable to compile a complete picture of the behaviour of a transducer using all the tools at our disposal, including engineering new approaches, for example, to account for environmental conditions.

An example of an electrical impedance spectrum for a Langevin ultrasonic transducer is shown in Figure 3.8, showing the series resonance frequency which

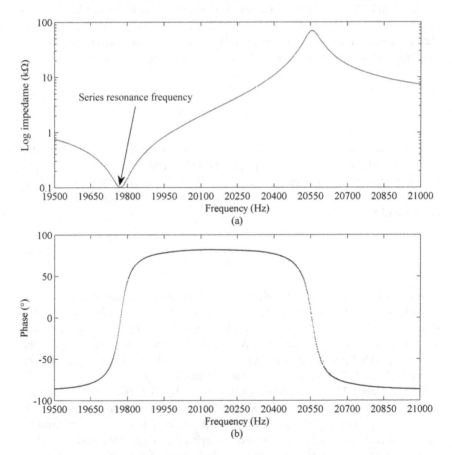

FIGURE 3.8   An example of an electrical impedance spectrum for a Langevin transducer, showing (a) the series resonance frequency and (b) the associated phase spectrum.

would typically be identified to align with the resonance of the device. The associated phase spectrum is also shown, illustrating the link with impedance. This series resonance will be linked to a physical mode of vibration for the transducer (and in this case it is the first longitudinal mode). In some cases, the electrical impedance spectra can provide instantaneous evidence of the condition or quality of a transducer. For example, it is common to measure double-resonances in reasonably close proximity, in the analysis of a Class V cymbal transducer (Bejarano et al., 2014). It is sometimes referred to as the double-peak phenomenon, and it typically indicates that one of the transducer's end-caps is vibrating with a different electromechanical resonance to the other. Therefore, it would be valid to assume the presence of asymmetry in terms of material properties and geometrical dimensions between the two end-caps and the epoxy resin bond layers in that case. An example of the double-peak phenomenon is illustrated later in Figure 4.8 in Chapter 4.

There are certainly other features of EIA which are important to the understanding of ultrasonic transducers, in terms of both the conventional configurations and those which incorporate advanced materials. For example, one can use EIA to measure and monitor the pre-stress on the stack of a Langevin transducer, by understanding the relationship between the applied torque on the central bolt and the charge output from the piezoelectric ceramic stack. The implications of this are that there can be a level of assurance on the consistency in the assembly of batches of transducers, and a threshold placed on the manufacture such that the physical integrity of the piezoelectric ceramic stack inside the transducer is maintained.

3.3.6.2 Laser Doppler Vibrometry

Optical measurements form an important part of the characterisation process for an ultrasonic device. In dynamic systems, it is usually desirable to study the vibrational behaviour of ultrasonic transducers, in terms of both modal response and harmonically at elevated levels of excitation. Optical measurements are valuable for studying the physical motions of different parts of a dynamic system, in the interests of tuning phenomena such as frequency response and the interaction with target materials and structures. Other phenomena of interest, for example, dynamic nonlinearity, can be expediently investigated through optical methods, and therefore can be important for assessing the practical operational limitations of an ultrasonic device. A common way that optical measurements are obtained is through LDV. The underlying principle of this system is interferometry, where two laser beams measure the phase difference between a laser beam focused on the surface of a test subject, and a second reference laser beam. These beams are combined by the LDV system, after where the dynamic response of the test subject can be mapped in the frequency domain. An example of an ultrasonic transducer undergoing a standard LDV measurement process is shown in Figure 3.9, which shows the focus of the laser beam from the laser Doppler vibrometer on the metallic surface of the transducer. This example is taken from a real experiment

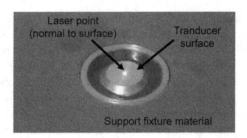

FIGURE 3.9 An ultrasonic transducer undergoing a standard laser Doppler vibrometry test for the measurement of mode shapes. The flat surface of the transducer in this case is in the order of 10 mm.

conducted on an aluminium flexural ultrasonic transducer, in the measurement of its modes of vibration.

In the example shown, the laser beam from the laser Doppler vibrometer would typically be focused normal to the vibrating target surface. The ultrasonic transducer would be excited using a suitable signal, commonly random, to ensure the frequency response functions in the range of interest are acquired. The support given to the ultrasonic transducer, or boundary condition, should be considered throughout the measurement, as this will influence the measurements. A typical laser Doppler vibrometer system makes use of the Doppler Effect through Equation (3.7), where there is a shift in the frequency of a wave relative to the frequency of the source, resulting in a measure of this phenomenon as f_d. In terms of the principle of interferometry, the laser Doppler vibrometer will make use of the combination of two laser beams (1 and 2 with r paths) in terms of their relative intensities I, as shown by Equation (3.8), extracted from Feeney (2014).

$$f_d = 2\left(\frac{c}{\lambda}\right) \tag{3.7}$$

$$I_{\text{total}} = I_1 + I_2 + 2\sqrt{I_1 I_2 \frac{\cos 2\pi (r_1 - r_2)}{\lambda}} \tag{3.8}$$

LDV can be used through two principal mechanisms in the characterisation of ultrasonic devices. The first is to capture the modal response of a test subject, in a process commonly referred to as experimental modal analysis. This can be considered as an alternative to using measurement approaches such as accelerometers. The principal disadvantage of an accelerometer is that it can typically necessitate placing undesirable mass on to a test subject, thereby directly biasing the validity of the modal response measurement. Second, accelerometers tend to require sufficient geometrical space, and this is not always possible. Many of the classes of ultrasonic transducer discussed in this book are of geometrical and mass scales such that accelerometers are not a viable option for the acquisition of high-quality measurements of modal response. Further to this, the resolution capabilities offered by modern LDV systems mean that it is a highly attractive option, particularly for the increasingly complex ultrasonic devices now being innovated. The other principal mechanism is the measurement of harmonic response. The key difference between the two approaches is the excitation provided to the test subject, and the parameters surrounding data collection. For experimental modal analysis, the objective is typically to measure the resonant modes in a particular frequency band, or that which is within the operational limit of the vibrometer, and it is usually undertaken with low excitation voltage levels and with a random signal. Harmonic analysis is typically undertaken to acquire amplitude responses as functions of frequency, for example, relating to specific excitation frequencies and voltage levels provided to an ultrasonic transducer. The specific parameters required for each measurement depend on the device under study.

Laser Doppler vibrometers can typically be acquired in single point or full-field scanning varieties. There are three-dimensional instruments available, which comprise three laser beams that capture dynamic response in three principal directions, allowing for a three-dimensional map of vibrational response to be collected. There are also vibrometers which are tailored to the measurement of rotational dynamics, and there are microscope-based vibrometers best suited to the study of micro-electromechanical systems (MEMS)-based devices. It is common that the single-point and scanning varieties allow measurements in a single plane, for example, in that normal to the direction of the laser beam emanating from the head of the laser Doppler vibrometer. In any case, the selection of laser Doppler vibrometer primarily depends on access and availability, given they are highly specialised instruments which can incur significant cost. Should there be a variety available to the engineer, it is advisable to consider what is desired from the optical measurements. Modal responses are more expediently captured using three-dimensional or scanning laser Doppler vibrometers, whereas single point vibrometers are highly suited to harmonic analyses.

There are several standard practices which can be followed in the proper use of LDV, to obtain high-quality optical measurements for a particular target application. These are briefly summarised in Figure 3.10, and a typical optical vibrometry scan for a common class of ultrasonic transducer, the flexural ultrasonic transducer, is depicted in Figure 3.11.

In general, it is strongly advised to record all parameters used in LDV studies such that measurements can be repeated, and so modal responses of interest can be readily identified with reliability. Through proper consideration of these factors, high-quality optical characterisation of ultrasonic devices can be undertaken, towards more complicated configurations for conducting such experiments on adaptive ultrasonic devices and those which incorporate materials necessitating a range of external environmental conditions.

Practice	Comments
Boundary Conditions	Careful consideration must be given to where the transducer is supported. It is common to utilise the nodal points or planes on a device to minimise the influence on the resonant mode. Nodal points and planes can be defined using finite element methods.
Expediency	The test subject should be configured so all measurement locations are accessible without needing to move the transducer. Surfaces of interest must be suitably reflective to ensure high-quality measurement. It is hence vital to first conduct a trial run.
Noise	Noise sources should be identified and removed, preventing bias or undesirable influence on the modal or harmonic measurements. Laser Doppler vibrometers are highly sensitive instruments, and so mitigating unwanted external sources of noise is important.

FIGURE 3.10 An overview of key considerations for laser Doppler vibrometry.

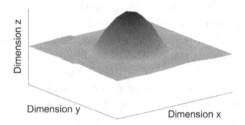

FIGURE 3.11 A typical mode shape measurement from a laser Doppler vibrometer, showing a flexural ultrasonic transducer vibrating at its fundamental resonance frequency. Here, the peak in the centre is the vibration of the front plate of the transducer.

3.3.6.3 Acoustic Microphone or Hydrophone Measurement

Transducers which are designed for the sensing of acoustic and ultrasonic waves can be characterised using the methods described thus far in this chapter. Dynamic features including resonance frequency and bandwidth are common to all acoustic or ultrasonic transducers, where EIA and LDV are both valuable and indispensable methods for measurement. In terms of sensing, there can be differences in approach depending on whether one is measuring ultrasound in a gas such as air, or in a liquid, such as water. A wide range of acoustic microphones are available, often tailored for measurement in air. From the commercial perspective, these microphones are generally capacitive systems. This means that the capacitance of the microphone itself is configured to change as a function of the waves impacting on the receiver component of the system, often a membrane or plate. There are also piezoelectric-based variants available. Conversely, a hydrophone can generally be considered as equivalent to a microphone, with the major difference being that it is employed for underwater measurement. Hydrophones are often piezoelectric, incorporating a piezoelectric element at the core of their structure with a tuned resonance frequency and bandwidth. These hydrophones are widely used in sonar technology, and there is a wealth of literature available on the design, fabrication, and operation of hydrophones for sonar and other measurement applications, with key texts relating to this field including Sherman and Butler (2007) and Stansfield and Elliott (2017). The principal reason a distinction should be made between microphones and hydrophones is due to acoustic impedance matching. A transducer whose structure is matched to air, in terms of acoustic impedance, will not work well in water, and vice versa. This is because the acoustic impedances of air and water are sufficiently different to reduce the efficiency if the transducers are not used in their intended environments.

If the discussion of (ultra)sound wave sensing is limited to the microphone in air for the purposes of this section, then the first step in the approach to recording high-quality measurements is similar to the advice given for other techniques. Effectively, the key dynamic characteristics of interest should first be determined. In general, for an ultrasonic sensor undergoing dynamic characterisation using an acoustic microphone, the selection of a suitable acoustic microphone is important. The resonance characteristics of the ultrasonic sensor should be understood, such that a reliable measurement can be made. Acoustic microphones tend to be complemented by a certificate of calibration, outlining the specification of the system and the criteria for accurate measurement. The acoustic microphone should be calibrated across a range of frequencies which encompass the resonance frequency of the ultrasonic sensor to be characterised. Second, an acoustic microphone can be used to obtain radiation patterns and directivity information for an ultrasonic sensor. By setting the propagation direction of the ultrasonic transducer along a particular axis, and then configuring the acoustic microphone to be rotated around the sensor in a circumferential path, a radiation pattern and directivity for a given coordinate system can be determined. An example of a radiation pattern measurement for a flexural ultrasonic transducer obtained using a typical experimental setup is illustrated in Figure 3.12 (Feeney et al., 2018b).

As shown in the sample data in Figure 3.12, there are a few features of interest of which an engineer developing ultrasonic transducers should be aware. The first is that a suitably wide angle of measurement should be used to capture the features of interest. This is especially important for higher order modes with features of interest, such as the data shown in Figure 3.12(b). It should also be ensured that the voltage level recorded is accurate, and that there is no significant interference or acoustic reflections that distort the data. Finally, there can be sidelobes in radiation patterns, and this can happen because of the geometrical features of a

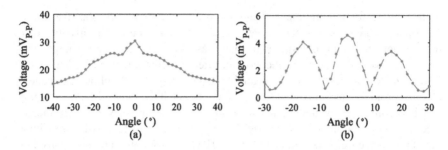

FIGURE 3.12 Example radiation pattern measurement using an acoustic microphone, allowing directivity of an ultrasonic sensor to be measured. Here (a) shows the radiation pattern of a fundamental vibration mode, and (b) shows that for a higher order mode. Reprinted (adapted) from the work of Feeney et al. (2018b), under the CC-BY 4.0 licence.

transducer and the system in which it operates, causing responses in the far-field which are not in the principal direction of wave propagation. Sidelobes generally result in lower efficiency and acoustical interference and are generally a mechanism by which energy in the system is wasted. Nevertheless, sidelobes are often unavoidable, though the size of the ultrasonic transducer and the characteristics of the principal or main radiating beam can both influence the nature of the sidelobes present in the radiation pattern.

Here for reference, the concepts of near-field and far-field for an ultrasonic sensor can be clarified. The near-field region designates the location in proximity to the radiating surface of an ultrasonic transducer, often characterised by a lower stability in the propagation of ultrasound waves. This area is often referred to as the Fresnel region, governed by a Fresnel distance (r_f) shown by Equation (3.9) and extracted using the work of Selvan and Janaswamy (2017).

$$r_f = \frac{d^2}{\lambda} \tag{3.9}$$

In this equation, the term d refers to the radius of the ultrasound source, and the λ parameter is the wavelength of ultrasound. This equation yields a useful marker for the transition point between near-field and far-field zones. Contrary to the Fresnel region, the far-field, or Fraunhofer, region constitutes characteristically stable and uniform wave propagation behaviour, typically divergent in nature, where the wave propagation is sometimes regarded as cone-like. The qualitative difference between the Fresnel and Fraunhofer regions are illustrated in Figure 3.13, and these can be experimentally verified with suitable laboratory instruments.

The underlying mathematics regarding Fresnel and Fraunhofer regions, and the propagation of ultrasound waves through different media, represent areas of significant and detailed research, such as Hyun and Dahl (2020). Although it is out of the scope of the book to provide that level of detail here, understanding what to look for in ultrasound sensor characterisation and the techniques available is

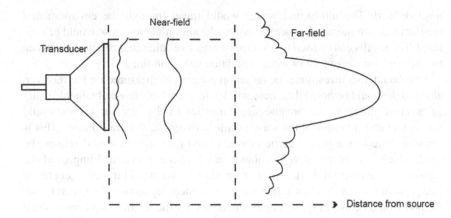

FIGURE 3.13   Qualitative representations of the near-field and far-field regions, broadly attributable to the Fresnel and Fraunhofer zones, respectively. Importantly, the waves typically diverge in the far-field, with less variation in intensity.

important. The configuration of the measurement or characterisation environment is critical, removing any sources of reflection which will bias or influence a measurement. There are a range of anechoic chambers available which can be used, but if correctly implemented and distributed, commercially available foams are known to be sufficient to mitigate the effects of reflections and unwanted sound sources.

Further to the use of acoustic microphones and hydrophones, there are alternative approaches to the characterisation of an ultrasonic transducer. The final method which this chapter addresses is the use of a receiver sensor to capture the dynamic characteristics of a transducer which is transmitting ultrasound.

3.3.6.4   Receiver Sensors

It is often desirable to capture the wave propagation behaviour from an ultrasonic transducer, in this case, usually a sensor, by using a secondary sensor as a receiver. This is a commonly applied technique for characterising sensor transducers such as the flexural ultrasonic transducer. It can be referred to as pitch-catch and is also undertaken in many industrial applications, such as flow measurement where ultrasound waves are transmitted across flow paths to enable time of flight to be determined. In general, it is usual to engineer a secondary receiver sensor as close to the physical properties of the transmitter, or sensor under study. In general, the dynamics of the transmission and receiver sensors should be as close as possible to enable reliable and high-value measurement. This reinforces the need for a high-quality and robust transducer fabrication method.

In this characterisation approach, it is advisable to position two sensors facing one another at a suitable distance of separation. The transmitter sensor would be used to generate a specific pulse of ultrasound waves at a given excitation voltage, number of cycles, and frequency which could be the resonance for the voltage level defined. The ultrasound waves would travel through the environmental medium to reach the receiver sensor, where the amplitude response would be captured. The results can be used to produce an effective calibrated condition between two sensors for a given set of inputs and boundary conditions.

One could also investigate the radiation patterns as measured on the receiver, although it should be noted that these will be different from those obtainable using an acoustic microphone, because the dynamics of the receiver sensor would become a factor in terms of the wave profile between the two transducers. This is primarily due to the piezoelectric effect, should piezoelectric-based sensors be used, which is a commutative phenomenon where the mechanical impact of the ultrasound waves from the transmitter on the receiver would thereby generate a charge in the receiver, thus influencing the system dynamics. In general, this method would be useful for setting a benchmark for the relationship between two ultrasonic transducers, or sensors, which could then be applied to subsequent fabricated sensors.

It should be noted that although the results from a dynamic characterisation process are often considered holistically, this requires some care. The series resonance frequency identified through EIA will not likely precisely match that measured using an optical method like LDV. Similarly, the resonance characteristics of a transducer will likely again be different should an acoustic method such as through a microphone be employed. The reasons for this are that the boundary conditions applied to the subject transducer under study in each case will not likely be consistent, the excitation conditions applied to the transducer will also be different (and this is especially important for any device incorporating a piezoelectric ceramic material), and the nature of ultrasonic vibrations are not the same if one considers mechanical responses versus electrical, or those which are acoustic and travelling through a medium such as air. The principal lesson from this is that each dynamic characterisation technique serves a particular purpose, where each can individually reveal a significant amount of information about a system or transducer. The techniques should therefore all be used together as part of a toolbox to learn more about the system, understand ways in which the transducer design can be optimised, and how best to employ the ultrasonic transducer under study for a given application.

3.3.6.5   A Short Note on Dynamic Nonlinearity

Prior to introducing a brief case study on the design and manufacture of an ultrasonic transducer, a brief overview of considerations around the phenomenon of dynamic nonlinearity is provided here. Dynamic nonlinearity is a broad aspect of dynamics which has been given considerable attention in recent years, with significant depth across a multitude of applications. It is not possible to account for all the achievements in this field for this book. However, there are several considerations regarding the dynamic characterisation of piezoelectric ultrasonic transducers, of which engineers should be aware.

First, the origins of nonlinear dynamic behaviours in the context of ultrasonic transducers can be clarified. What one would consider a real system, such as an ultrasonic transducer designed for a particular application and therefore subject to a given set of input or excitation conditions, can generally be mathematically approximated as a linear system in many cases. This means that a given parameter, such as an input voltage to the transducer, will produce a proportionate output amplitude. This approximation can be useful for a variety of characterisation purposes, but it is not particularly advantageous for accurately understanding or representing the dynamic response of a system in practice. Dynamic responses, including for those systems incorporating piezoelectric materials, typically exhibit amplitude responses which are not in direct proportion with their input conditions. For example, the propensity of the resonance frequency of piezoelectric power ultrasonic transducers to reduce as the excitation voltage is raised beyond a specific threshold, known as nonlinear softening, has been widely reported (Mathieson et al., 2013; Liu et al., 2024b), and there are also behaviours where the resonance frequency will nonlinearly increase as the excitation voltage is raised, known as nonlinear hardening. The transition from an approximate linear response to nonlinear behaviour is important to consider, in addition to where such characteristic responses manifest from and how they might be mitigated through transducer design. There are interesting design implications for adaptive ultrasonic transducers, including those incorporating shape memory alloys, and this is discussed in Chapter 4. Nonlinear behaviours in dynamic systems can arise from several sources, but three which are key for the subject matter of this book include nonlinearities attributable to the following classifications:

1. **Material.** There are many inherently nonlinear characteristics associated with materials from which piezoelectric ultrasonic transducers are fabricated. For example, the resonant properties of piezoelectric materials themselves can change as the input voltage is raised (thus influencing the temperature of the material), and Nitinol exhibits a nonlinear relationship between stress and strain.
2. **Geometry.** The size and shape of component materials in relation to other parts of a dynamic system will influence the response of that system.
3. **Boundary Condition.** An ultrasonic transducer can be supported or fixed in place in many ways, and in each case, any additional parameter or dimension on a boundary condition as applied to a dynamic system increases its complexity. This can range from clamped features to bolted joints.

The principal point to consider here is that through a relatively straightforward modification to the classical linear approximation of a dynamic system, the governing mathematical relationship for which is shown by Equation (3.10), the nonlinear response of a dynamic system can be expressed using the relationship shown by Equation (3.11).

$$\ddot{x} + \mu\omega_0\dot{x} + \omega_0 x = A\cos\Omega t \qquad (3.10)$$

$$\ddot{x} + \mu\omega_0\dot{x} + \omega_0 x + \eta x^3 = A\cos\Omega t \qquad (3.11)$$

Here, the addition of a cubic term in the relationship equates to a nonlinear stiffness, and restructures the linear approximation as a nonlinear equation, and is a form of the Duffing equation (Mathieson et al., 2013; Feeney et al., 2019a). The implication of this cubic term in understanding the dynamic characteristics of piezoelectric ultrasonic transducers is that experimentally observable phenomena can be represented dependent on the sign of this cubic term. If negative, then the system can be said to exhibit nonlinear softening, where in qualitative terms the amplitude response curve, as a function of frequency, bends to the left (or lower natural frequencies). Conversely, a positive cubic term is indicative of nonlinear hardening, where the amplitude response curve bends towards the right, or higher natural frequencies. The typical forms of Duffing-type responses are depicted in Figure 3.14, extracted from the work of Mathieson et al. (2015b).

There are further implications of the softening and hardening nonlinear behaviours which are of relevance for the configurations of piezoelectric ultrasonic transducer discussed in this book. The first is how to accurately capture nonlinear dynamic behaviour through experiment, and ways in which one can clearly identify the sources of the nonlinearity. The second is that if there were sufficient prominence in the nonlinearity, and if the system is operated at an excitation frequency in proximity to a region of instability and at a sufficient amplitude, there can be a *jump* phenomenon where the amplitude is unstable and switches between higher and lower levels of amplitude (Mathieson et al., 2013). Care should therefore be taken regarding the frequency at which a piezoelectric ultrasonic transducer is operated, given its nonlinear dynamic response and amplitude output, both of which should be characterised prior to use.

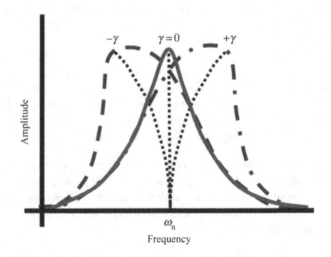

FIGURE 3.14 Characteristic nonlinear dynamic responses, showing softening ($-\gamma$) and hardening ($+\gamma$) behaviours, in comparison to a linear response ($\gamma = 0$). Reprinted from the work of Mathieson et al. (2015b), under the CC-BY 4.0 licence.

As a final comment in this chapter regarding the consideration of dynamic nonlinearity and the methodical approaches for understanding and mitigating the influences, it should be acknowledged that this is a phenomenon which will affect the results associated with every system incorporating piezoelectric ultrasonic devices. The consideration and characterisation of these behaviours is important to capture the full capabilities of the system, and to map the ways in which the system can be optimised. The final section of this chapter will collate the material discussed thus far as required into a concise case study on the design, fabrication, and characterisation of a flexural ultrasonic transducer.

3.4 CASE STUDY: A FLEXURAL ULTRASONIC TRANSDUCER

In this final section of the chapter, a typical piezoelectric ultrasonic transducer is selected to demonstrate the design, fabrication, and characterisation principles associated with practical ultrasonic transducer applications. A flexural ultrasonic transducer is chosen here as a test case for the complete design from first principles towards a functional device, in part due to its relative simplicity in terms of the required component materials, but also because it is a class of transducer which has thus far not received significant attention in the literature. Therefore, this is an opportunity to learn more about a novel form of ultrasonic transducer which is becoming more widely used in a variety of industrial measurement applications. Some of the material here is taken from Kang et al. (2020), under the CC-BY 4.0 licence.

It is appropriate here to revisit the schematic of the transducer development process shown in Figure 3.2, where it can be assumed that Steps 1 and 2 have already taken place. Regarding Step 1, *defining the operational parameters*, and in terms of the flexural ultrasonic transducer, thought should be given to the environment in which the device is operating and the measurements that are important to capture. In this case, the flexural ultrasonic transducer would typically be operated in a gas, so an air-coupled environment can reasonably be assumed, and one for generic proximity measurement. A flexural ultrasonic transducer is an ideal choice of transducer for this environment, consistent with the approach for Step 2, *deciding the class of transducer required*, given that this class of transducer is particularly well suited to measurement in gaseous media such as air without the requirement of a matching layer. Further to this, we can then look to answering some of the questions presented in Figure 3.4, here shown in Figure 3.15.

Regarding **boundary conditions**, given the nature of how the flexural ultrasonic transducer works and its operating mechanism whereby the dynamics of the transducer are dominated by the vibration behaviour of the membrane or plate, a typical cylindrical form of housing would be suitable, and it can be assumed that the transducer would be fixed around its circumferential external surface during operation. This should be ensured for the finite element model.

The desired **dynamics** of the flexural ultrasonic transducer directly inform the materials which should be incorporated into the design, and their physical dimensions and properties. A sensible target resonance **frequency** for this transducer is in the range of 40 kHz–100 kHz, based on commercial equivalents and

Consideration	Initial Criteria
Boundary Conditions	Transducer has fixed proximity to target, usually at far-field distances. Assumed uniform clamping around transducer housing.
Dynamics	No specific requirement on bandwidth. Fundamental axisymmetric mode of vibration (here, the (0,0) mode). No coupling with other modes of vibration.
Environment	Air, in situ. Ambient atmospheric pressure and approximately room temperature. No other environmental considerations at this stage.
Frequency	Target operating frequency in the region of 40 kHz to 50 kHz, typical for commercial flexural ultrasonic transducers.
Excitation	The transducer will be a sensor, so lower power operation is expected, and therefore low voltage excitation (for example, a limit of 10 V_{P-P}).
Production	One transducer can be fabricated, to be operated in pulse-echo mode.

FIGURE 3.15   The key considerations from Figure 3.4 answered for this case study.

the drive towards high-frequency applications (Kang et al., 2020). The transducer is typically operated in a fundamental axisymmetric mode of vibration because these modes are generally used to optimally generate relatively high amplitudes of vibration for only a few volts of **excitation**. Furthermore, there should be no significant risk of coupling to proximate modes of vibration, and in particular asymmetric modes. If required, finite element analysis can be used to assess this.

Given the **environment** is set as ambient air, there will not be any significant restrictions on the type of material used for the membrane or the piezoelectric ceramic disc. For example, we would not need to consider impacts of higher temperatures which might de-pole a particular piezoelectric material or degrade a candidate epoxy resin. Further to this, a standard insulating epoxy resin can also be selected to bond these components together, and so a conventional type of epoxy resin such as Araldite® could be used. Finally, regarding the requirement for one sensor, there does not need to be specifications put in place for the consistent fabrication of multiple matched transducers. However, if the **production** stage did require multiple sensors, then a fabrication rig able to deliver a rapid manufacturing process, for example, the capacity to cure the epoxy resin bond layers within several transducer configurations, would be advantageous. Thus, in summary:

- The assumed operating environment is air at ambient room pressure and temperature.
- A conventional form of flexural ultrasonic transducer can be used.
- The transducer is fixed around the circumferential external surface of its housing.
- Operation at its fundamental axisymmetric mode of vibration, which is optimal for the transducer.

It is known that the geometrical and material properties of the cap of the flexural ultrasonic transducer are fundamental to the transducer's dynamic behaviour. The vibration response of this class of transducer is commonly regarded as equivalent to that of an edge-clamped plate since the dynamics of the transducer are largely dominated by the vibration of the plate. It is therefore expedient to identify the transverse displacement of the plate w through the differential equation shown through Equation (3.12), defined using characteristics equations (Feeney et al., 2019a; Kang et al., 2020).

$$DV^4 w(\bar{x},t) + \rho h \frac{\partial^2 w(\bar{x},t)}{\partial t^2} = 0 \qquad (3.12)$$

Here, the plate rigidity D is given in terms of the radius x and the thickness h. The density of the cap material is ρ. It should be noted that this rigidity magnitude can itself be defined in terms of the elastic constants of the plate, specifically Young's modulus E and Poisson's ratio v, as shown in Equation (3.13), via (Feeney et al., 2019a).

$$D = \frac{Eh^3}{12(1-v^2)} \qquad (3.13)$$

Using these relationships, an expression for the resonance frequency f for the edge-clamped plate can be generated, using a Bessel function formation which is related to the mode shape, and where $x = a$ is a boundary condition ensuring that the radius of the cap plate is a. This relationship is shown by Equation (3.14), via (Feeney et al., 2019a).

$$f = \frac{1}{2\pi} \left(\frac{\lambda}{a}\right)^2 \sqrt{\frac{D}{\rho h}} \qquad (3.14)$$

If we consider titanium as an example candidate plate material, since it is mechanically robust and is suitable for engineering ultrasonic transducers with target operating frequencies in the kHz range. Furthermore, its elastic properties are generally consistent with the parameters required for the fabrication of flexural ultrasonic transducers at a size equivalent to currently available commercial devices. Using the characteristic equations given as (3.12) through to (3.14), then f can be predicted as **23.23 kHz** for the first axisymmetric mode of vibration, or the (0,0) mode, and **90.44 kHz** for the second axisymmetric mode of vibration, or the (1,0) mode. These calculations were made using assumed material properties for titanium of $\rho = 4510$ kg/m^3, $\nu = 0.32$, $E = 116$ GPa, and geometrical properties of $h = 0.25$ mm and $a = x = 5.20$ mm.

There are a few points to raise regarding the predicted resonance frequencies of the transducer, as provided above. The analogy of the edge-clamped plate for the flexural ultrasonic transducer has been known and utilised for several years, and it has provided a useful foundation for transducer design. The design principles here can be extended to other configurations of transducer. Nevertheless, the validity of this approach greatly depends on the chosen materials for the transducer design, and the available control on the fabricated transducer components and the associated manufacturing and assembly processes. If this calculation is undertaken for a commercially available flexural ultrasonic transducer, such as those reported in published works (Dixon et al., 2017; Feeney et al., 2018a, 2018b), then this approximation is relatively close. There is an acceptable level of control assumed in the manufacturing process of commercial sensors, as they are fabricated within control limits (it is not unusual to find resonance frequencies stated around ± 1 kHz for 40 kHz transducers).

However, through repeated calculation using the given equations, and modifying the parameters, it is plain for the reader to see how sensitive the calculated resonance frequency is to even minor adjustments in the magnitudes of some of the parameters. For example, in the case of the f for the fundamental (0,0) axisymmetric mode of vibration, reducing the plate thickness by 0.1 mm results in a drop of 9.29 kHz in the estimate of f, to 13.94 kHz. In manufacturing control terms, the difference between 0.25 mm and 0.15 mm in these scales may not seem significant. Therefore, care should be taken when using this initial estimation of f, and in the physical manufacture of the transducers, where control of the fabrication process is important. The final point to make regarding the mathematical prediction of the transducer's resonance frequencies is that this estimation is specifically derived from the theory of plate vibrations (Leissa, 1969). This theory, whilst valuable, is not strictly a true equivalent to a real transducer. It is not a finite element simulation, but rather an expedient way to estimate the approximate magnitude of f. There are other mathematical analogue approaches available for different classes of transducer, and each can be highly useful in setting the parameters of a more comprehensive finite element model which can be used to generate data for comparison with experimental measurements.

The next step is to use this information as a guide and build the parameters into a finite element simulation, to allow prediction of transducer behaviour.

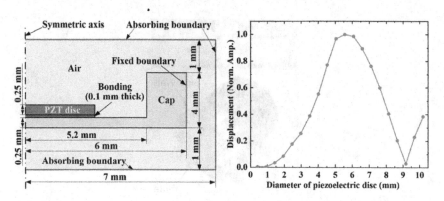

FIGURE 3.16 The design schematic for the flexural ultrasonic transducer, suitable for a finite element model (left), with the simulated place displacements as a function of piezoelectric disc diameter (right). Reprinted (adapted) from the work of Kang et al. (2020), under a CC-BY 4.0 licence.

A valuable aspect of a high-quality finite element model is that it allows the simulation of conditions which might be difficult to apply in practice, for example, in multiphysics software such as COMSOL Multiphysics® (COMSOL, Inc.), and allows an investigation into the sensitivity of a range of dynamic characteristics in response to changes in certain physical parameters, including geometry and elastic properties. Here, *OnScale Solve* is used to build a two-dimensional finite element model, where the external surfaces of the transducer cap are assumed to be rigidly fixed, forming the desired boundary condition consistent with the assumptions made at the start of the case study. Then, the relevant material properties for titanium are included in the model, along with those for the epoxy resin bond layer, via a thin layer in between the piezoelectric ceramic and plate layers. A schematic for the transducer design is shown in Figure 3.16, alongside simulated transducer plate displacements for different diameters of piezoelectric ceramic disc, to optimise the design of the transducer for the output performance (Kang et al., 2020).

The piezoelectric ceramic disc is chosen as a standard form of PZT-5H, a soft piezoelectric ceramic suited to sensing applications, and incorporates a wrapped electrode for ease of access to both electrodes after fabrication. The wrapped electrode means that an electrical voltage can be applied to the piezoelectric ceramic disc whilst one face is bonded to the transducer membrane. Once the transducer model is generated, its dynamic response can be simulated for a given mode of vibration. Using the simulation parameters, the transducer materials can be gathered, and the transducer fabricated, as shown in Figure 3.17.

Before any representative experimental characterisation results are shown, there should be clarifications around the manufacturing and assembly process. It should also be noted that the information provided here is intended to be indicative of what one might expect in the design and characterisation process. Readers are encouraged to review the relevant papers in the scientific literature which are

FIGURE 3.17 The fabricated flexural ultrasonic transducer. Reprinted (adapted) from the work of Kang et al. (2020), under a CC-BY 4.0 licence.

openly available and accessible, for more detailed accounts of transducer design and characterisation. The transducer, as shown in Figure 3.17, is composed of a cap machined from a single specimen of titanium. This can be a rod, or in some cases from suitably thick sheet material. However, it is possible to approach fabrication using alternative methods. For example, the plate and the side wall could be fabricated separately, and using different materials, to facilitate greater precision and control in the plate properties. This can also be useful to ensure a perpendicular interface between the plate and the side wall. It can be difficult to ensure this by other conventional manufacturing means, but it does have a significant impact on transducer dynamics. Using such a method, the plate can be rigidly bonded into the structure and cured using applied pressure from a bespoke curing fixture, in a manner similar to how the piezoelectric ceramic disc can be centrally bonded to the plate. Once these steps are complete, suitable wires can be carefully soldered to the piezoelectric ceramic disc, and then a protective backing attached to the rear of the device to protect the internal components from damage.

In the case of the titanium flexural ultrasonic transducer that is the subject of this case study, after fabrication it is ready to be characterised using the EIA, LDV, and acoustic microphone measurement methods outlined in this chapter. Typical characterisation results for these techniques are depicted in Figures 3.18 and 3.19 for reference (Kang et al., 2020), where the electrical impedance spectrum and mode shape of the first axisymmetric mode of vibration are shown in Figure 3.18, and the amplitude–time data with the radiation pattern measurement in the far-field in the first axisymmetric mode of vibration are shown in Figure 3.19. The results shown in Figure 3.19a were captured using an acoustic microphone.

Regarding the EIA, it is evident that the local series resonances indicate the vibration modes of the transducer. As shown previously in this chapter, EIA is not a useful technique for providing information regarding the physical mode shapes of a transducer, but it is highly effective for providing information regarding resonance frequencies and their proximity to one another. It is also a valuable technique for rapid evaluation and correlation with mathematical modelling that has

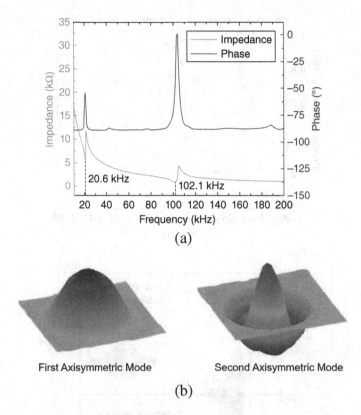

(a)

First Axisymmetric Mode   Second Axisymmetric Mode

(b)

FIGURE 3.18 The electrical and optical measurements of the titanium flexural ultrasonic transducer, showing (a) the electrical impedance spectrum and (b) the optically measured axisymmetric mode shapes via laser Doppler vibrometry, which are 20.6 kHz and 102.1 kHz for the first and second axisymmetric modes, respectively. Reprinted (adapted) from the work of Kang et al. (2020), under a CC-BY 4.0 licence.

been undertaken. Notably, the resonance frequencies of the two modes of interest, at 20.6 kHz and 102.1 kHz, are close to the resonance frequencies of those modes calculated using the characteristic equations earlier in this section (23.23 kHz and 90.44 kHz, respectively). There is a greater discrepancy for the mathematical prediction for the higher order vibration mode, likely because the mode involves more complex vibration motion with an increased quantity of nodes in the mode shape. Therefore, this can produce a greater discrepancy between the mathematical model and the experimental result. The measurements associated with ultrasound wave propagation as shown in Figure 3.19 can be used to demonstrate the operational capacity of the transducer for a practical measurement application.

The result in Figure 3.19(a) shows the radiation pattern in the far-field for the transmitter in the second axisymmetric mode of vibration. Specifically, this result can be directly compared with the vibration mode shape measured using LDV as shown in Figure 3.18(b). The link between the physical motion of the vibrating

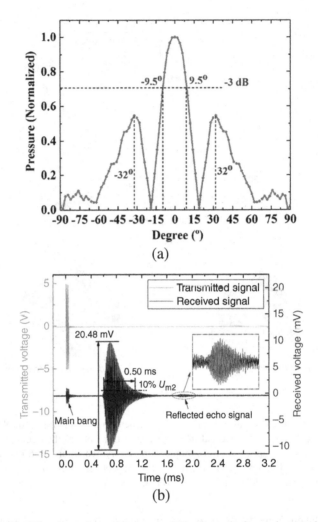

FIGURE 3.19 The dynamic performance of the flexural ultrasonic transducer in air, showing (a) the radiation pattern in the far-field measured with an acoustic microphone, and (b) the amplitude–time response at resonance using a matched receiver transducer. Reprinted (adapted) from the work of Kang et al. (2020), under a CC-BY 4.0 licence.

plate and the propagating ultrasound waves is evident. The dynamic performance of the transducer as both a transmitter (generator) and receiver (detector) of ultrasound is shown via the data in Figure 3.19(b). Here, a matched pair of transducers is used in a pitch-catch configuration to show that a sufficiently strong signal can be generated and detected using the transducers, exhibiting potential for practical industrial measurement applications.

As a final note regarding this case study, although much of the information can be found in the published scientific literature, the aim of this section has been to highlight the key steps in the design, fabrication, and characterisation of an ultrasonic transducer. A similar approach can be applied to the production of other types of ultrasonic transducer, though there will be inevitable minor differences associated with each.

3.5 SUMMARY

This chapter has aimed to outline the key considerations for the conception, design, fabrication, and experimental characterisation of an ultrasonic transducer. This general overview has been largely limited to piezoelectric-based configurations, and those classes of transducer commonly utilised in both industrial and medical applications for sensing and power ultrasonic processes.

The first part of this chapter discussed some of the more practical considerations for transducer design, including the questions an engineer can ask to develop the required understanding of the system with transducer, advice for how to select the optimal transducer configuration for a target application, and which modelling strategy can be followed to expedite the experimental aspects of transducer development. From this, a suitable manufacturing process can be planned before the necessary experimental characterisation is undertaken.

This chapter has also included some advice, from both the scientific literature and experience, in the selection on piezoelectric materials for ultrasonic devices. The principal lesson from this is that consultation with industry experts is key, and that this is central to the successful fabrication of a high-performance ultrasonic transducer. Whilst bulk piezoelectric ceramics remain popular choices of active material for ultrasonic transducers, the material landscape is rapidly changing, particularly with the advent of textured ceramics and high-precision fabrication methods to synthesise intricate shapes and layers of ceramic materials on small scales, such as through tape casting. Thorough reviews of available approaches should be undertaken, consistent with the objectives of the transducer development or the target application. The choice of material has a direct influence on the mathematical modelling approach employed, which itself influences the manufacturing method which is most suitable for the ultrasonic transducer. Proper planning is therefore imperative to ensure robust devices are designed and fabricated, able to be consistently replicated in future manufacturing processes.

The chapter concluded with a brief case study of the design and fabrication of a bespoke flexural ultrasonic transducer using key information from the scientific literature, consistent in terms of configuration with one which could in theory be readily integrated with a shape memory material to constitute an adaptive ultrasonic device. The intention of this section was to illustrate how key principles discussed earlier in the chapter could be utilised to design, fabricate, and characterise a functional piezoelectric based sensor. Example experimental data was presented as supportive material, and it was demonstrated that by using a combination of analytical modelling and finite element analysis, a sensor configuration for ultrasonic proximity measurement can be readily engineered using the information presented in this chapter.

4 The Material Challenge

4.1 SCOPE

The research activity regarding the specific classes of smart or advanced material of interest here for acoustic and ultrasonic devices, shape memory alloys and acoustic metamaterials, experienced a shift around the start of the 21st century. For example, the tuneable frequency flextensional cymbal transducer fabricated using the shape memory alloy Nitinol was demonstrated in 2000 (Meyer, Jr. and Newnham, 2000). Here, it was shown that the frequency response of the ultrasonic transducer could be controlled by manipulating the unique phase transformational characteristics of the vibrating Nitinol end-caps, thus constituting the first real detailed example of an adaptive ultrasonic transducer fabricated from a shape memory material. It should be noted though that the concept was introduced almost 10 years earlier, as reported in the work of Newnham (Newnham, 1992). Nevertheless, there have been a series of broad but significant explorations in several research directions from 2000 to the present day, based on these initial studies, with other forms of ultrasonic transducer now under investigation. The scope of discussion and analysis for this chapter can be clarified, including definitions of smart or advanced materials. According to Kennedy et al., advanced materials are (Kennedy et al., 2019):

> 'Materials that are specifically engineered to exhibit novel or enhanced properties that confer superior performance relative to conventional materials'.

It is arguable that this classification of material is limited to those which are engineered by us, and with tailored properties designed to generate a particular response in a system. The term *advanced material* is broad, and like so many technical concepts, it would not be practical to encompass every key aspect in one text. Advanced materials are often considered alongside smart materials, where, for example, in 2022, Athanassiadis et al. published a comprehensive review of smart materials and systems specific to the field of ultrasound, arguing that smart devices are those which can switch their states, adaptively responding to the influences of external stimuli or triggers (Athanassiadis et al., 2021). This theme of adaptiveness is very much a primary focus of this book, and therefore the content of this chapter will explore the materials we can use to achieve this. In terms of ultrasound, Athanassiadis et al. presented a schematic illustrating the principal capabilities of smart materials, as enabled by ultrasound, as shown in Figure 4.1 (Athanassiadis et al., 2021).

Here, characteristic behaviours can be designated across the triggering and control of biological or chemical processes; the reconfiguration of materials, systems, and structures; complex sensing and actuation; robotics; and motion control.

 DOI: 10.1201/9781003324126-4

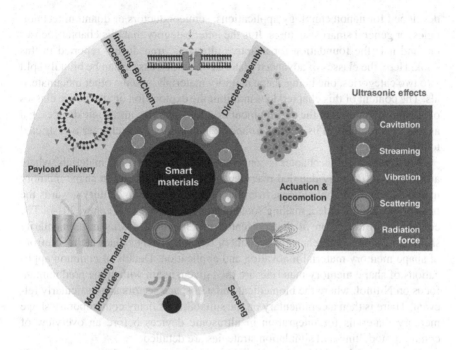

Initiating Processes · BioChem. · Directed assembly · Ultrasonic effects · Cavitation · Streaming · Vibration · Scattering · Radiation force · Payload delivery · Smart materials · Actuation & locomotion · Modulating material properties · Sensing

FIGURE 4.1 The capabilities of smart materials in the field of ultrasound. Reprinted (adapted) from the work of Athanassiadis et al. (2021), under the CC-BY 4.0 licence.

Athanassiadis et al. make the case for such capabilities to be enabled in smart systems and structures via ultrasound phenomena including cavitation and structural vibrations. These factors are undoubtedly key in relation to how shape memory materials and acoustic metamaterials will find success in ultrasonic technologies, but the argument can also be made that ultrasonic technologies already provide the ideal platform for broadening the application space for these classes of smart material. From this point onwards, the shape memory alloys and acoustic metamaterials discussed in this chapter will be considered as smart or advanced materials. It also must be clarified that in many cases, shape memory alloys and acoustic metamaterials will generally play different roles in acoustics and ultrasonics, and therefore some of their characteristic behaviours and typical applications will be detailed in this chapter. Technology in this space remains at relatively low readiness, and so some of the applications discussed are suggested as potentials for future exploitation of these classes of material.

In general, there are many combinations of elements and minerals which constitute engineered or advanced materials, and so the focus of this discipline must be narrower than those which relate to the description included above. For example, piezoelectric ceramic materials are generally regarded as advanced materials given their piezoelectric effect which can be reasonably classified as an *enhanced property*. Advanced materials could constitute those which are bioengineered, or developed for nanotechnology applications, semiconductors or quantum technologies, or general smart structures. It is the latter category that this chapter focuses on, and it is the foundation for adaptive ultrasonic transducers reported in this book. Here, the classes of advanced material under discussion can be broadly split into two categories, one being shape memory materials, and the other metamaterials. The content of this chapter does not limit the exploration of alternative classes of advanced material in the development of future adaptive ultrasonic devices, but as this field remains in its relative infancy, the content of this chapter is restricted to these two classes.

The first part of this chapter is concerned with shape memory materials, where a brief but sufficiently detailed overview of the developments in shape memory materials and their applications, from the middle of the 20th century towards the present day, are included, making some detailed reference to (Feeney, 2014). The content is an extension of that presented in Chapter 2. Focus is momentarily shifted to a more comprehensive account of Nitinol, given its place at the forefront of shape memory material innovation and application. Details of common applications of shape memory materials are then given, again with some predominant focus on Nitinol, where the biomedical and aerospace sectors are particularly relevant. There is then a commentary on the suitability of many contemporary shape memory materials for integration in ultrasonic devices before an overview of common modelling and simulation strategies are detailed.

The scope of this chapter then proceeds with a more directed approach towards the thermal and mechanical characterisation of Nitinol, including key considerations relating to practical integration of the material into ultrasonic devices, after which an account of device design advice is provided. The chapter then moves to a commentary on the opportunities afforded by metamaterials and metastructures in ultrasonic devices. The general goal of this chapter is therefore to provide some fundamental underpinning knowledge for future device designers to take forward into new applications.

4.2   SHAPE MEMORY MATERIALS

4.2.1   A BRIEF HISTORICAL BACKGROUND

The emergence of shape memory alloys is often regarded as taking place in 1932 with the demonstration of shape memory in an alloy of gold and cadmium, Au-Cd, by Arne Ölander (Gil and Planell, 1998). Here, the alloy was deformed in a cool state before being heated, where it was observed to recover its original set configuration, demonstrating what was in effect a *memory*. Not long after that, around 1938, Greninger and Mooradian were able to show a similar phenomenon in an alloy of copper and zinc, Cu-Zn, which culminated in the fortuitous but famous discovery of Nitinol at the Naval Ordnance Laboratory in the early 1960s (Thompson, 2000). The focus of this chapter will shift more prominently towards Nitinol in the subsequent section, where it will be treated in much greater detail, but it is important to demonstrate that despite the prevalence and popularity of

Nitinol today, it was an inadvertent discovery and was not a feature of the first array of shape memory materials investigated. A useful overview of the timeline of shape memory alloys and their approximate discoveries and developments is depicted in the work of Alipour et al. for those interested in further reading on the topic (Alipour et al., 2022).

In 1949, Kurdjumov and Khandros were responsible for publishing their understanding of how a reversible phase transformation was possible (Kumar and Lagoudas, 2008), paving the way for future shape memory alloy development. Their work focused on the thermoelastic behaviour present in copper-based alloys, predominantly Cu–Zn and Cu–Al. With an improved understanding of reversible phase transformations, the principle of the martensitic phase transformation became more understood, which then stimulated research into other combinations of chemical elements and the development of other alloys, as detailed in the work of Alipour et al. (2022). For example, in the 1950s, there were observations of superelastic, or pseudoelastic, behaviour in alloys such as Cu–Zn, Cu–Al–Ni, and In–Tl (Gil and Planell, 1998), and which is a behaviour which will be covered in more detail later in this chapter. Progressing from Ölander's initial observation of a memory-like behaviour in Au–Cd in 1932, Chang and Read recognised the shape memory effect present in Au–Cd with 47.5 at% Cd in 1951 (Kastner, 2012). Progressing from this, William Buehler et al.'s famous discovery followed in the early 1960s.

4.2.2 Unlocking Nitinol's Potential in Ultrasonics

From here, we can shift our focus towards Nitinol as the key composition of shape memory alloy for integration with ultrasonic devices. As a binary alloy, it has arguably been the most popular and successful of the shape memory materials to this day, closely followed by copper-based alloys which are particularly well suited to environments of elevated temperature. Nitinol is a binary alloy of nickel and titanium, and its origins can be traced back to the late 1950s and into the 1960s, when the fascinating mechanical and thermal characteristics of the material were recognised at the Naval Ordnance Laboratory (Thompson, 2000). Given its binary composition, it was given the name Nitinol in recognition of where it was developed and innovated by engineers, notably including William Buehler and Frederick Wang (Jani et al., 2014). The discovery of the principal material characteristic of interest, the shape memory effect, took place effectively by accident. The story is that the research at the Naval Ordnance Laboratory was focusing on the development of new materials which could exhibit improved resistance to impact and fatigue at higher levels of temperature (Kumar and Lagoudas, 2008). Nitinol was found to be a particularly desirable candidate, but at a meeting within the Naval Ordnance Laboratory, one of the attendees was noted to have applied heat to a deformed sample of Nitinol, only to witness the sample of material recover to its original set shape (Kumar and Lagoudas, 2008). From this point forward, efforts have been made to realise the potential benefits of this material, to investigate the influence of hot and cold working, different compositions of

Nitinol and the effect of various alloying elements, and optimal machining and processing conditions, and has been the subject of many investigations to improve and broaden their application space (Jani et al., 2014).

Nitinol is non-ferrous by its nature and composition (Newnham, 1992), and it is binary because its composition is generally equiatomic between nickel and titanium elements. As referred to above, various alloying elements can be included in the composition to enhance its material properties and transformational behaviours, for example, the temperatures at which the phase microstructure shifts from martensite to austenite, or vice versa. The material properties and transformational response of Nitinol has been found to be highly sensitive to the chemical composition of the alloy, and significant research has progressed for many years to optimise its response for various applications. Nitinol generally exhibits excellent resistance to corrosion because it generates a protective oxide layer on its surface which is rich in titanium, and it is also a biocompatible material. Due to this, it is commonly applied in the fabrication of biomedical stents (Stöckel, 1995, Stoeckel 2000; Duerig et al., 1999, 2003; Pelton et al., 2000; Stoeckel et al., 2004; Liu et al., 2008; Bach et al., 2013), exploiting its superelastic behaviour, which will be outlined in a little more detail in this chapter. Nitinol alloys are known to be able to endure relatively high levels of stress (Hall, 1994; Shaw 2008), which makes them particularly attractive for power ultrasonic applications, and they can recover comparatively high levels of strain introduced into the material. They are also known for their relatively favourable damping characteristics (Van Humbeeck, 1999), which has made them a point of interest for aerospace applications. The transformation temperatures of Nitinol can also be tuned up to around 100°C (Chen et al., 2009), though they are still typically utilised at temperatures below this, for example, around room temperature in the case of some applications including biomedical stents utilising superelastic responses. As an extension to the historical development of shape memory alloys, but Nitinol more specifically, it is important to note that alloying elements can make a notable difference in the performance and characteristics of the material. Some of this will be outlined in more detail later in the chapter. However, from a developmental perspective and to briefly illustrate how the field has developed through the years, soon after the discovery of Nitinol, it was found in 1965 that the transformation temperatures of the material could be reduced, and hence controlled, via the addition of iron or cobalt (Kumar and Lagoudas, 2008). This was a significant development because it showed that there was scope to engineer Nitinol for a wide range of applications across science and engineering.

Nitinol was successfully employed in the early 1970s in the development of the Cryofit technology, for aerospace and defence applications (Saadat et al., 2002; Kumar and Lagoudas, 2008). Niobium was then used as an alloying addition to Nitinol in the 1980s to achieve a sufficiently large temperature hysteresis in the material (Kumar and Lagoudas, 2008), meaning that it could effectively behave like Cryofit but not require transportation at very low temperatures, for example, in liquid nitrogen, like it did formerly. Around the decades of the 1970s and 1980s, significant exciting research and development took place in the creation of high-temperature shape memory alloys, such as those with transformation temperatures greater than 100°C (Kumar and Lagoudas, 2008). Although this part of the chapter focuses on Nitinol in particular, such alloys are typically based on titanium alloyed with materials such as platinum, gold, or palladium. However, research into Nitinol continued, and in the late 1970s, it was demonstrated by Mercier that by introducing copper into the Nitinol alloy, it was possible to decrease the stress hysteresis inherent in the material, but generally preserve the transformation temperatures. It would take several years, into the late 1990s, for Miyazaki to show that copper is key to introducing improvements to the fatigue life of Nitinol (Kumar and Lagoudas, 2008).

From the 1970s to the 1990s, great efforts were made to engineer Nitinol for exploitation in the biomedical field, and it is now a popular material choice for stents. Furthermore, Nitinol is now a material of choice for dentistry and other medical devices, in addition to a selection of industrial applications (Saadat et al., 2002). Notably, Andreasen and Hilleman were able to demonstrate Nitinol archwires for orthodontics as early as the 1970s (Stöckel, 1995; Thompson, 2000). Latterly, Nitinol has been utilised in various applications including as air conditioner vent materials, in acoustic control, switches, composite materials, miniature actuators, and cabling connectors (Saadat et al., 2002; Kumar and Lagoudas, 2008). In general, Nitinol is known to be more expensive in terms of fabrication compared to some copper-based alloys such as Cu–Zn–Al, but it remains an extremely popular choice of material, especially in terms of commercial applications (Chen et al., 2009). It can be made biocompatible, it has relatively high corrosion resistance, and it is relatively straightforward to manufacture in simple configurations such as wires, sheets, and ribbons, making it attractive for actuation applications. The continued popularity of Nitinol and the development of alternative forms is partly due to the progressions in advanced manufacturing and measurement capabilities (Pelton et al., 2000; Shaw, 2008; Alapati et al., 2009; Jalaeefar and Asgarian, 2013; Fernandes et al., 2013). Experimental measurement techniques such as differential scanning calorimetry, X-ray diffraction, and dynamic mechanical analysis are invaluable for monitoring and measuring the properties and physical response of Nitinol and other shape memory alloys. They are fast becoming vital as key steps in the design and manufacture of adaptive ultrasonic devices, as will be illustrated in this chapter. The final key point to raise here is that there remain some unknowns about shape memory alloys and their responses when embedded in ultrasonic devices, and through continued development, robust standardised approaches for device design and development can be established.

Nitinol can be complicated to manufacture, and it is commercially available in forms such as wires, plates, sheets, and ribbons. However, the emergence of more reliable advanced manufacturing and the use of techniques such as electrical discharge machining or diamond-tip machining means that more intricate and tailored designs can now be fabricated. More recently, progress in the additive manufacturing of shape memory alloys, including Nitinol, has meant that even more complex designs can now be realised. These approaches will also be discussed in this chapter, but it is important to note that significant progress is

required to qualify or validate Nitinol components for ultrasonic transducers manufactured using these approaches. One reason for this is that additively manufactured Nitinol has not been trialled to any significant extent in ultrasonic transducers,
and another is because there have been some concerns surrounding factors relating to the longevity or suitability of the material for a vibrating system, including
its fatigue life (Jani et al., 2014). Other common applications from a commercial
point of view in recent years have included (effectively) damage-resistant spec-
tacle frames, arch-wires in orthodontics, and as strain-recoverable or -resistant
components in minimally invasive devices for surgery (Kumar and Lagoudas,
2008). The main material property of interest for these applications is the superelastic effect, one of the two interesting material behaviours which arises from the
underlying mechanisms surrounding its phase transformations.

Since much of this chapter, at least from the materials perspective, is focused
on Nitinol, it is necessary to give an illustrative overview of how Nitinol's phase
transformations can be produced, and the associated ways in which various
mechanical and material properties can change. At its core, the phase microstructure of Nitinol can exist as either a symmetric, cubic B2 austenite at relatively
high temperatures, dependent on the alloy composition (Paula et al., 2004), or a
lesser-ordered martensitic microstructure, which is monoclinic and B19' in nature
(McNaney et al., 2003). These B terms can be compared with the relevant
Strukturbericht designations, which are informative of how each crystal structure
appears. Examples of these crystal structures are illustrated in Figure 4.2, extracted
from the work of Shimoga et al. (2021).

In general, a lower-temperature martensitic phase microstructure in Nitinol
does not strictly have to conform to a monoclinic configuration, for example,

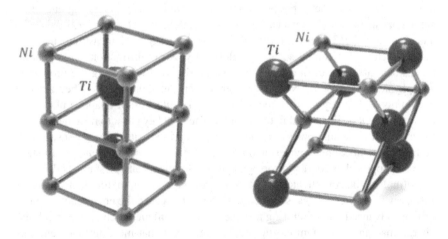

FIGURE 4.2 Examples of the major crystal structures present in Nitinol, showing (left)
the configuration of the material's phase microstructure in its B2 austenite phase, and (right)
the crystal structure of the material in its B19' martensite phase. Reprinted (adapted) from
the work of Shimoga et al. (2021), under the CC-BY 4.0 licence.

where it can be triclinic or orthorhombal (Shaw, 2008). The dependency of this is primarily on the alloy composition. Given that Nitinol possesses at least two distinct phase microstructures which can be tuned and set, or triggered using a particular stimulus, there are hence a range of material properties whose characteristics can also be tailored or tuned, depending on the phase microstructure. For example, austenite and martensite phase microstructures in Nitinol can exhibit different elastic properties, electrical resistivities, and crystal structures. Properties such as the coefficient of thermal expansion and the thermal conductivity of the material are also both dependent on the phase microstructure. The potential of this for realising adaptive ultrasonic devices is therefore significant.

The most prominent and important material property for adaptive ultrasonic devices is arguably the elastic modulus because this is inherent to the resonant properties of an ultrasonic transducer. There are different classifications of modulus for a material, and here we will focus on what we would regard as Young's modulus, that which is the dependent on the relationship between stress and strain in a material. Young's modulus of Nitinol in its martensitic phase is conventionally regarded as within the 30–40 GPa region, or thereabouts, whereas the modulus when the material is austenitic is around 60–90 GPa. It is not uncommon for commercial suppliers to state such ranges typical for Young's moduli of Nitinol in these phase microstructures. It is evident that these range specifications are sizeable in terms of their tolerance or uncertainty, and this underlines a major challenge of integrating Nitinol, and materials with similar properties, into ultrasonic devices. The material properties can be difficult to control and equally challenging to accurately measure. It has even been observed that the use of resonant ultrasound spectroscopy can still yield results that are difficult to link to a realistic understanding of modulus (Thomasová et al., 2015; Grabec et al., 2021). Principally, the material properties of Nitinol are highly dependent on the chemical composition of the alloy, and the mechanical processing conditions to which the material is subjected. For example, complex and demanding machining and processing is generally required for introducing threaded features into a Nitinol component, with significant potential physical influences on the material and its properties.

In general, given the stated Young's modulus ranges for both martensitic and austenitic phase microstructures, it is evident that the material will stiffen as its temperature is raised. This is the contrary or opposite behaviour to that of many more conventional metallic materials commonly adopted in ultrasonic devices. The accepted terminology here is that the reverse transformation is that which refers to the transformation from martensite to austenite (McNaney et al., 2003), whereas the forward transformation is that which is associated with the switch from austenite to martensite. The triggers for these phase microstructure transformations are either temperature, which will be prevalent for much of the work reported in this book, or stress (Thompson, 2000), which is a common stimulus in the underlying mechanisms behind superelasticity. There is a phase transformation in Nitinol which arises from the mechanism of shear lattice distortion (Kumar and Lagoudas, 2008). This means that transformation does not occur via atomic diffusion, and it is a key factor in how each phase can be recovered via either of the triggers of temperature or stress (loading). Phase transformations are generally regarded as martensitic transformations, and it is common to see it written as such in the academic or discipline-specific literature. Key to how Nitinol can be integrated into an ultrasonic device, martensite can exist in two major forms. The first is twinned martensite, where the orientations in the microstructure, or the *variants*, are accommodated, and the second is detwinned martensite, where a specific orientation is dominant. In the former, there is no tangible shape change observable in the material, whereas in the latter, there is. This could be achieved by applying a suitable deformation load in the martensitic phase. Finally, the transformation behaviour is governed by a series of transformation temperatures, which are those denoting the transitions from key stages in the phase microstructure of the material to the others. They are classified as either start X_S or finish X_F temperatures, where X denotes the phase microstructure under consideration. Therefore, each phase microstructure possesses a start and finish temperature, but it should be noted that the temperature at which a phase microstructure forms on a heating cycle may not correspond with the temperature required to form it once again on a cooling cycle. This is because Nitinol, like many shape memory materials, is highly hysteretic in nature. This will be revisited later in this chapter.

Prior to proceeding with an overview of the specific material behaviours associated with Nitinol, it is important to note here that there is often an intermediate phase microstructure present in Nitinol, referred to as the R-phase (Ling and Kaplow, 1981; Huang and Liu, 2001). This is a rhombohedral-type microstructure, and it can appear in either one or both of the heating or cooling cycles, as applied to a sample of material. It therefore also possesses associated X_S and X_F temperatures, and great care is needed to not confuse the R-phase with martensite. The R-phase is strictly a distortion of austenite (Shaw, 2008), and it is sometimes regarded as the *pre-martensite* phase, but it can generally be considered as a martensitic-type phase because it will exhibit shape memory and superelastic behaviours (Duerig, 2012). Superelasticity and shape memory are outlined in detail in Section 4.2.3, but in general a key distinction for the R-phase is that its superelastic abilities are minor compared to that of the principal austenitic phase in Nitinol, being in the order of 0.5% in terms of strain distortion. Importantly, in the context of ultrasonic devices, the R-phase can be induced by stress (Miyazaki and Otsuka, 1986), and therefore it can manifest in either twinned or detwinned configurations. This means that martensite and R-phase could be triggered through stress, or temperature, and so it is important to be aware of which phase microstructure is being exploited for a particular case. To illustrate the differences between the three principal phase microstructures in Nitinol, Figure 4.3 depicts the crystal structure of each principal phase microstructure of Nitinol, and how they change through the transformational processes of thermal shape memory and superelasticity (Guo et al., 2013).

The thermomechanics at the foundation of superelasticity and thermal shape memory will be outlined in greater detail in Section 4.2.3, but the key points here are that these transformational behaviours are hysteretic, and they can generate

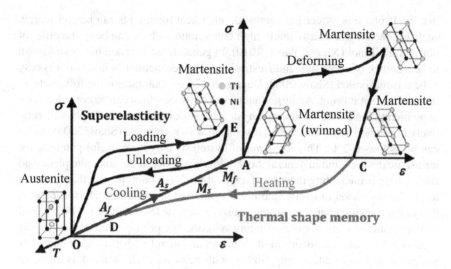

FIGURE 4.3 Superelasticity, shape memory, and the schematic differences between martensitic and austenitic Nitinol. Reprinted from the CIRP Annals – Manufacturing Technology, Vol. 62, Guo et al., Machinability and surface integrity of Nitinol shape memory alloy, Page 83, Copyright (2013), with permission from Elsevier.

different phase microstructures in the material. On a further point regarding the R-phase, it is evident that it can exist in the cooling of a sample of Nitinol from its austenitic microstructure to martensite, but it is also possible to observe it in the transition from martensite to austenite, and such instances can be found in the scientific literature (Huang and Liu, 2001; Shaw, 2008). There are reasons why the R-phase may exist in one path and not the other, but in general a direct transformation from austenite to martensite is common. It is much more common to experience the R-phase in alloys which are richer in nickel (Shaw, 2008), or those which have undergone a range of heat treatments or cold working, or processing in certain solutions (Uchil et al., 1998). Furthermore, cycling the material through temperature can also trigger the emergence of the R-phase in the material, and for these reasons, suitable thermoanalytical measurement is strongly advised. This will be outlined later in this chapter.

In terms of ultrasonic devices, it might be appropriate to encourage the emergence of the R-phase. One reason might be that the difference in transformation temperatures required to transition between a martensitic microstructure and austenite can be optimally reduced, thus providing a more practical method of switching modulus of the material, and hence resonance frequency of an ultrasonic device. This could theoretically be achieved by designing appropriate ageing or solution treatments of the material which promote the generation of nickel-titanium precipitates in the material (Eaton-Evans et al., 2008). This would likely result in the generation of the R-phase, and hence a shorter temperature window by which a phase transformation could be achieved. However, should it be desired that the R-phase is reduced or removed, other heat treatments can be considered, such as annealing at a sufficiently high temperature, which can be in the order of 600°C for Nitinol (Xu and Wang, 2010). In general, the R-phase has been known to be a source of confusion and frustration for experimentalists because it is easy to be misinterpreted (Shaw, 2008; Duerig, 2012), for example, as the full martensite phase, when it is not. An important piece of advice which can be considered for experimentation is that the thermal hysteresis associated with the R-phase is relatively low, when compared to that associated with martensite (Shaw, 2008), and it can be as low as 2°C. This low level of hysteresis may have useful potential for ultrasonic devices, and in general the relationship between the austenite phase and the R-phase is interesting in terms of application. It is a combination that is known to yield a high level of cyclic stability if compared in terms of performance with that of the transition between austenite and martensite (Shaw, 2008), which would be important for vibratory or oscillating systems. The principal reason behind this is that the R-phase is a distortion of austenite, and therefore has a higher degree of structural and kinematic compatibility with austenite, which itself is cubic in structure. It has even been recorded that the R-phase could be induced via stress from the martensite phase (Duerig, 2006), which would have interesting applications for adaptive ultrasonic devices. Therefore, the R-phase, which has for a long time been a source of frustration, may be a desirable solution or opportunity for exploitation in ultrasonic devices, especially for systems facilitating or accommodating lower levels of stress and strain. The utilisation of the R-phase will be revisited later in this chapter.

4.2.3 CONTROLLING TRANSFORMATION BEHAVIOUR

In this section, some of the key interesting behaviours present in Nitinol will be introduced, with perspectives of how these properties can be integrated into an ultrasonic device and exploited. In general, these properties are *superelasticity* and the *shape memory effect*. Thus far, they have been referred to in the contexts of the material properties of Nitinol or in terms of its application, but it is important for a transducer designer to understand what each of these behaviours can do and the underlying mechanisms. The reader is directed to Figure 4.3 for schematic context alongside the descriptions provided in this section.

4.2.3.1 Superelasticity

In general, a Nitinol alloy is superelastic if it exhibits a highly reversible or recoverable response to a stress applied to the material, and it is an extremely elastic phenomenon appearing when the Nitinol exists above its final austenitic transformation temperature A_F. Its elastic response can be in the order of 10 times that of stainless steel, as a guideline (Stöckel, 1998), and so it is particularly advantageous for engineering robust and recoverable devices. It is common to see superelasticity referred to as pseudoelasticity, particularly by materials scientists (Kumar and Lagoudas, 2008), given its underlying mechanism of shifting domain boundaries. Nitinol is extremely mechanically hysteretic (Stöckel, 1998), and so it is possible to recover the strain imposed in a sample of material from an applied stress at a lower stress magnitude than the stress applied to produce the strain in the first instance.

Superelasticity can be thought of as an isothermal process, principally because it does not require an external temperature trigger to produce. Instead, it exists when the phase microstructure of the material is austenitic in nature, which transforms to martensite under the application of a sufficiently high stress (Shaw, 2008; Kumar and Lagoudas, 2008). Given that there is a dependency of transformation temperatures in a shape memory material such as Nitinol on stress magnitude, it follows that there can be a reorientation of the phase microstructure from stiffer austenite to relatively more compliant martensite, above a specific stress threshold and providing the temperature of the material is appropriate for the austenitic phase microstructure to be present. It is also for these reasons that the action of superelasticity is effectively instantaneous, at least to an observer, because stress-induced martensite would typically be unstable under the noted conditions, in those required with respect to temperature for the generation of the austenitic phase microstructure. Both the thermal and mechanical responses of Nitinol are hysteretic (Poncet, 2000), and therefore the relationship between stress and strain can be relatively complex. The underlying mechanism of superelasticity, and how the phase microstructure reorients in the Nitinol alloy, is illustrated in Figure 4.3, which can be considered alongside the deformation mechanism for a more conventional Hookean elastic material (Pelton et al., 2000).

It is evident from the deformation response mechanism illustrated in Figure 4.3 that under the application of sufficiently high stress to induce a phase transformation from austenite to martensite, there is a progressive change introduced into the crystal structure of Nitinol. In practical terms, there may hence be localised regions of superelastic response in a sample of material, and this is significant for how stress concentrations in vibrating components of a Nitinol-based ultrasonic device would be influenced. For example, one might think of particularly high-stress regions of an end-effector fabricated from Nitinol to exist in a different microstructure phase compared to a region of the end-effector in a lower stress condition.

The basic overview of how the superelastic response is generated can therefore be defined. If the phase microstructure of Nitinol is complete austenite, then as a stress is applied, the microstructure progressively reorients to martensite, so long as the applied stress is sufficiently high. It is common to encounter references to upper and lower plateaus, and these can typically refer to the loading and unloading stages, respectively (Schlun et al., 2011). It is from these regions that a quantification of the mechanical hysteresis in Nitinol can be assessed and understood. If we were to translate this into a real-world practical example, one might align the compression condition for a biomedical stent fabricated from superelastic Nitinol to be dependent on the upper plateau stress, because that is relevant for a loading condition. If the upper plateau stress is higher, then a higher force would be required to compress and store the biomedical stent in the desired configuration. This has relevance to an adaptive ultrasonic device, because in the first instance one would need to understand the condition of the Nitinol to be embedded in the device, and secondly it would be important to measure and analyse the material behaviours of the Nitinol prior to integration. One of these behaviours might be the plateau stresses, or the criteria for generating the superelastic response in the material. It would then follow that the lower plateau stress would relate to the release of the Nitinol biomedical stent and the full recovery of the austenitic phase microstructure. Up to this point, the discussion has been framed around both stress and temperature, and there is a governing relationship which is commonly utilised in the field, and it is a form of the Clausius–Clapeyron equation, shown by Equation (4.1) and extracted from Pelton et al. (2000).

$$\frac{d\sigma}{dT} = -\frac{\Delta H}{T\varepsilon_0} \qquad (4.1)$$

Here, the formula shown by Equation (4.1) provides a relationship between stress σ and temperature T for stable phase microstructures of austenite and martensite in the Nitinol. The transformational strain ε_0 is shown to be part of the dependency, in addition to the latent heat shown by the H term. One of its key applications is to provide an estimation of the reversible strain that is achievable, considering the stress magnitudes as functions of strain (McKelvey and Ritchie, 2000; Pelton et al., 2000). Since both H and T are in general constants in the circumstances here, the relationship shown by Equation (4.1) is a measure of stress-induced martensite and how much stress is required to achieve this in a specimen of Nitinol (Duerig, 2012). This has a significant bearing on how a transducer designer might consider superelasticity, or the emergence of the phenomenon in practice, through links with finite element simulation data in parallel during the transducer design process. There may be an amplitude or excitation condition it would be advisory to avoid, should the superelastic effect be unwanted in operation of a transducer. Of course, should the superelastic phenomenon be desired, then it would be important to ensure that the final austenitic transformation temperature A_F for the Nitinol is below the operating temperature of the adaptive ultrasonic device in practice, considering the influence of different stress levels, because this would ensure the material is superelastic in operational conditions. As an illustrative representation of the relationship between stress and temperature in Nitinol, the schematic shown in Figure 4.3 demonstrates how the applied stress σ at a given temperature T can generate different phase microstructures in the material (Guo et al., 2013), since stress and temperature both influence the transformation of the material's phase microstructure from one condition to another.

Considering the information shown in Figure 4.3, the initial four key transformation temperatures relating the start and finish of the martensite and austenite phase microstructures, respectively (thus M_S, M_F, A_S, and A_F), and neglecting the R-phase for the purposes of this example, are shown first in a zero-stress state by their locations on the horizontal abscissa. The transformation temperatures are dependent on stress (Kumar and Lagoudas, 2008), and therefore by inducing a specific stress level in the Nitinol material, different phase microstructures can be generated that will have tangible and possibly significant impacts on the dynamic performance of an adaptive ultrasonic device. It is also evident that there are many possible loading paths which can lead to the generation of stress-induced martensite, and this is without considering the influence of the R-phase within this. Finally, it should be noted that there is also a temperature at which the superelastic phenomenon would generally not appear, often referred to as the martensite desist temperature, or M_D. It has been noted that this can be in the order of around 50°C above A_F (Stöckel, 1998), and it is another feature of importance to transducer designers hoping to exploit the superelastic phenomenon.

4.2.3.2   The Shape Memory Effect

Arguably the most dominant and certainly the most well-known and familiar physical phenomenon present in a shape memory alloy such as Nitinol is the shape memory effect. It is often just referred to through its acronym, as the SME, and in general it refers to the ability of a shape memory alloy to recover its original set shape or configuration through a temperature stimulus (Jani et al., 2014). A sample of Nitinol alloy can be fabricated to exist as different physical shapes based on the phase microstructure induced in the material at a given time. This is not mandatory, and it is common to just exploit the changes in mechanical properties such as Young's modulus, as much of the scientific literature has demonstrated. A general illustration of the shape memory effect can be considered in the following list:

1. Nitinol begins in its austenitic phase microstructure condition, above A_F.
2. The sample is then cooled to generate pure martensite, below M_F, where in this condition it can be regarded as twinned martensite.
3. A stress is then applied to the material to introduce a physical shape change to the sample, deforming it out of its set configuration.
4. If a temperature gradient is then applied to the sample, such that its temperature is raised past A_F, then the original shape of the sample will be recovered. This is the basic operation of the shape memory effect, and it typically occurs instantaneously, in effect with reference to an observer.

There are many representations of the shape memory effect in the scientific literature, including relating both twinned and detwinned martensite into the physical response mechanisms, but an illustrative example of the shape memory effect in action is shown in Figure 4.4, on a sample of Nitinol wire.

In the example shown in Figure 4.4, the Nitinol wire is first twisted out of shape when it exists in its martensitic condition, or below its A_F transformation temperature. An electrical current is then passed through the wire, where its temperature gradient can be monitored using an infrared thermal imaging camera. It should be noted that the switching between shapes can be achieved repeatedly should it be desired, and if the environmental and operational conditions are appropriate, providing that the deformation conditions applied to the material do not introduce any

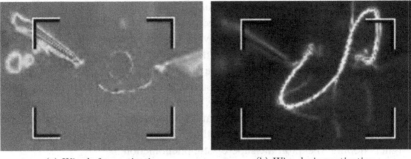

(a) Wire before activation   (b) Wire during activation

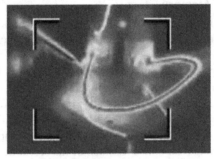

(c) Wire after activation

FIGURE 4.4   The shape memory effect in a sample of Nitinol wire, showing the deformed wire in its lower temperature martensitic condition in (a), before being heated via an electric current through its transformation in (b), towards an austenitic microstructure in (c). Reprinted (adapted) from the work of Feeney (2014).

permanent effects which may disrupt or destroy it. Even though Nitinol is a highly robust and resilient material exhibiting relatively high strength, it does possess a plastic strain limit and so it will experience deformations after which the material would not be able to recover its parent shape in the austenitic phase microstructure (Kumar and Lagoudas, 2008). The ability of Nitinol to rapidly transform between a range of phase microstructures with different mechanical properties is one of the major reasons why it is being investigated for its potential in adaptive ultrasonics. It has already been used in transducers with remotely repairable features (Meyer, Jr. and Newnham, 2000), and given the stress and temperature dependencies, there is significant scope to further our understanding of how the shape memory effect can be activated and controlled through ultrasonic vibrations (Rubanik et al., 2008; Belyaev et al., 2014; Lesota et al., 2019). An example of this is the application of 20 kHz vibrations in a sample of Nitinol plate inserted into an ultrasonic waveguide (Lesota et al., 2019), where it was demonstrated that the shape memory effect could be generated for sufficiently low levels of residual strain. Another study reported in 2020 demonstrated the use of focused ultrasound to stimulate the phase transition in Nitinol wire, thus providing a non-invasive mechanism to trigger the shape memory effect via heating (Chipana et al., 2023). As an interesting supplementary example, a novel approach for utilising the phase transformation in Nitinol for industrial applications was reported by Jain et al. in the demonstration of a Nitinol sensor for two-phase (e.g., gas and liquid) flow detection (Jain et al., 2020). Here, a sensor in the form of a wire was conceived to take advantage of the change in electrical resistance with temperature in Nitinol. Although this does not necessarily specifically rely on the shape memory effect to work, it is an example of the practical implementation of this material, and there is scope for its application with the shape memory effect beyond this.

Furthermore, different shapes can be set in the various phase microstructures of Nitinol, one phenomenon of note being the two-way shape memory effect (Kumar and Lagoudas, 2008). Any shape setting and configuration of the shape memory effect generally requires thermomechanical training in the material. A particularly well-trialled approach to this is through repeated cycling of the temperature gradient, between hot and cold conditions to generate the desired phase microstructures in the material, whilst the Nitinol specimen is fixed in an intended shape. The combination of this mechanical loading and the thermal cycling constitutes a robust training regime for the material, where eventually the optimal path for reorientation within the phase microstructure becomes aligned with that to produce the desired shape recovery in the specimen, including with regards to the formation of favoured martensitic variants which become dominant in the microstructure.

4.2.3.3 Engineering Transformation Control

The main question which arises from all the material presented thus far on what Nitinol is and how its properties can be controlled is that how can this all translate into a practical adaptive ultrasonic device? There is significant scope for some intelligent and innovative engineering to create novel trigger mechanisms for switching the mechanical properties of the material across desired temperature ranges or stress thresholds, or even for some device concepts which generate a transformation or switching in the device performance in a different way altogether, using the material properties as a basis. Young's modulus has been a major focus thus far, but this is not the only property of Nitinol whose characteristics are dependent on the phase microstructure of the material, and this should be considered in future device design. In general, prior to the design of an adaptive ultrasonic device from the perspective of its dynamics and resonant properties, great care should be taken to ensure that the thermomechanical processing conditions are understood, including their influence on the transformation temperatures of the material. In addition to influences on the transformation temperatures, there are also associated influences on the mechanical characteristics (Kuhn and Jordan, 2002; Shaw, 2008). The transformation temperatures can be tuned through different approaches, and so understanding how these methods can be used to adapt the material is imperative and can be relatively time consuming. The main approach is to tailor the chemical composition (Pelton et al., 2003; Shaw, 2008), for example, where if there is a desire to raise the transformation temperatures then the quantity of nickel in the alloy could be reduced to achieve this (Kumar and Lagoudas, 2008). Transformation temperatures can also be raised by alloying the material with elements such as aluminium, copper, and manganese, usually in relatively small quantities (Chen et al., 2009). Of course, making changes to the chemical composition is not straightforward and requires a good working relationship with a manufacturer, if material is acquired on a commercial basis. The other important aspect of this is that in general, for Nitinol in its conventional commercially available forms, it would normally be expected that transformation temperatures can only be tuned up to around 110°C (Narayanan et al., 2007), though there are copper-based shape memory alloys available which can reach higher temperatures in this regard. In a practical sense, the tuning of temperatures below 0°C is perhaps not as critical, but temperatures for transformation have been known to be as low as around −200°C. The influence of very minor changes in nickel or titanium content can be significant to the transformation temperatures, where, for example, raising the nickel content of an alloy by 1% above 50 at% of nickel can result in a reduction of transformation temperatures in the order of 100°C (Shaw, 2008). This has serious implications for device design, and it is therefore strongly recommended that materials science experts and manufacturers are properly engaged within the design and development process. It has been noted that with the right expertise, it is possible to control the transformation temperatures of Nitinol to within ± 3°C of what is desired from a specification (Pelton et al., 2000). Given current manufacturing capabilities, tighter control may be possible, but this will likely be dependent on the configuration of the Nitinol specimen. Complex shapes require more processing steps including cold working, which affect transformation temperatures. It is also important to note that various mechanical properties of Nitinol, including strength, will change depending on the quantity of alloying elements in the material (Gallardo Fuentes et al., 2002), and so the definition of the material condition must be undertaken on a holistic basis for the device design and in anticipation of its mechanical and dynamic performance.

The other principal method of shifting the transformation temperatures is through thermal–mechanical processing. One of these is a cold working process of cyclic loading (Pelton et al., 2000; Shaw, 2008), and the other is an ageing heat treatment process, for example, in the 350–500°C temperature range. Cold working can be highly effective in controlling the transformation temperatures of a material, and it can also be necessary to configure the shape of a structure (Pelton et al., 2003). It is customary to introduce some post cold working heat treatment, for example, in the order of 600°C or higher, for the principal reason that it can lead to rapid work hardening. In terms of the processing of Nitinol, there has been some success in using techniques such as magnetron sputtering to deposit thin films of Nitinol (Shin et al., 2004; Adams et al., 2005; Martins et al., 2005), but there have been concerns regarding the reproducibility of these materials in terms of their transformation characteristics. It is generally regarded as a significant challenge in translating Nitinol into different configurations of transducer, and controlling the transformation temperatures with precision is a key factor in ensuring viability and quality.

FIGURE 4.5 Examples of the influence of annealing heat treatments on specimens of Nitinol end-caps intended for a Class V cymbal transducer, showing (left) an untreated specimen, (middle) an end-cap annealed for 1 h at 450°C in air with light surface abrasion, and (right) a Nitinol end-cap subjected to the same annealing conditions but for 2 h, with no abrasion of the surfaces to show the condition of the oxide layer. Reprinted (adapted) from the work of Feeney (2014).

There have also been reports of how thermomechanical cycling of Nitinol has been implemented to introduce changes into the hysteresis in the material (Kumar and Lagoudas, 2008), shifts in transformation temperature (Tadaki et al., 1987; Zarnetta et al., 2010; Song et al., 2013), and even the two-way shape memory effect (Kumar and Lagoudas, 2008). Many of these changes occur because dislocations or similar are introduced into the microstructure of the material, or where preferential modes of microstructure transformation are introduced. The key point from this is that a considered approach to alloy composition, cold working, and heat treatment can yield a Nitinol alloy wholly suited for a particular application or for an adaptive ultrasonic device of interest. The time-consuming aspect of this is the design of the material condition and the manufacturing process required. Examples of what different samples of Nitinol look like resulting from different annealing heat treatments are shown in Figure 4.5 (Feeney, 2014), where the oxide layers can be observed, especially in the case of the sample annealed for 2 h at 450°C in air. A transducer designer should consider the implications of this, for example, if other materials need to be adhered to the surface after annealing. This could include soldering or similar, where this would likely be more difficult in the presence of this oxide layer.

The summation of the material presented in this chapter thus far has been to inform the reader on shape memory alloys and the specifics of Nitinol and how it works in general. Most of the subsequent information presented in this chapter, before discussions on metamaterials and their role in adaptive ultrasonics are presented, relates to experimental aspects of incorporating Nitinol shape memory alloys into ultrasonic devices. Before we reach that point, a brief note on modelling and simulation relating to Nitinol transducers is included, as a pre-requisite step for the experimentally relevant characterisation and operational advice.

4.2.4 MODELLING AND SIMULATION CONSIDERATIONS

A brief overview of some of the common approaches in the modelling and simulation of ultrasonic transducers incorporating shape memory materials is included in this section. In general, the modelling and simulation is reported here from the perspective of finite element analysis. There are many detailed and comprehensive accounts available of the constitutive and analytical modelling of shape memory behaviour, particularly pertaining to the thermal and mechanical behaviours associated with Nitinol (Kumar and Lagoudas, 2008). It is not the intention of this book to replicate this, but it is important to be aware of the detailed literature available.

Finite element analysis software is widespread and prevalent, and it is of vital importance to modern ultrasonic transducer designers. In recent years, significant efforts have been made to progress the capabilities regarding the generation of finite element models capable of accurately representing shape memory alloys and their unique behaviours, notably the shape memory effect and superelasticity. The complexities surrounding the mechanical properties of Nitinol, and their measurement, are explored in more detail later in this chapter, but the central advice from the perspective of modelling and simulation is that given there are notable and distinct challenges in the measurement of a multitude of material and transformational properties associated with Nitinol, it should not be expected that the modelling and simulation of these materials and the devices into which they are integrated is any simpler.

Today, finite element software platforms such as COMSOL Multiphysics® (COMSOL, Inc.) possess embedded tools for simulating the physical response of shape memory alloys, including structural and thermal functions. More specifically, these tools tend to be focused on the simulation and calculation of stress as a function of strain for a specimen of material, approaching the nonlinear regime. The underlying modelling strategies will typically involve classical constitutive models including those famously developed by Lagoudas (Kumar and Lagoudas, 2008), and they will generally require several input properties to ensure a reliable prediction of stress–strain response is predicted. These may include a reference temperature, for example, relating to the environment in which the model is designed to replicate or represent, key material properties including density and Poisson's ratio, and Young's modulus of whichever phase microstructure is of interest. There may also be several more input properties required relating to the stress and strain thresholds.

It has also been common to develop user material subroutines for various finite element analysis software packages, including Abaqus/CAE (Wang and Yue, 2007; Vidal et al., 2008). In such cases, these subroutines, also known as UMAT subroutines, are effectively algorithms which govern the physical response of the Nitinol or shape memory alloy, across its thermomechanical, superelastic, and shape memory behaviours. Common examples of specimens in model form include springs and wires since they are widely used on a practical basis and in commercial applications. The key point associated with this is that it is advisable to first decide whether a finite element analysis model is required, and if so, which outputs are desired. It would be expected that for a designer of an ultrasonic transducer, a robust finite element analysis model would be vital in the interests of identifying the resonance frequency of a targeted operational mode shape. Since

Young's modulus is of vital importance to transducer design, but given it is also very difficult to measure (a point detailed further in this chapter), a viable route would be to estimate a sensible range of magnitudes for Young's modulus in each phase microstructure for Nitinol, and then build these into a series of models. A practical range of geometries for a Nitinol component to be integrated with the transducer design could then be identified, and once a prototype is fabricated, the finite element model could be effectively reverse engineered. It is arguable that an approach such as this, whilst not perfect, avoids the presupposition of too many parameters in the model, for the Nitinol shape memory alloy itself, such that there is limited utility and value in the model.

An approach which has been followed in the design of both Langevin type and flextensional class ultrasonic transducers incorporating complex shapes of Nitinol is through using estimates of Young's modulus for each phase microstructure of the material informed through a combination of dynamic measurements and mathematical modelling (Feeney, 2014; Liu et al., 2024b). In these cases, the modulus magnitude was determined to be the principal property of importance to the transducer designers, and best available estimates of Young's modulus were used from manufacturer data. In the models, the dynamic or resonance response could be adjusted and tuned through modification of the modulus magnitude, with the general assumption that other material properties and dimensions in the model were defined with an acceptable level of accuracy. There is evidence in the scientific literature that this has been a successful approach in the past (Meyer, Jr. and Newnham, 2000). The other aspect of simulating shape memory alloys is that they are highly nonlinear in their nature, and with a hysteretic response in terms of their performance as a function of temperature. Even in a singular phase microstructure, Nitinol does not exhibit a stable or constant magnitude of modulus with changing temperature (Duerig, 2012), and this reinforces the need for robust and reliable material property data where possible (Perry and Labossiere, 2004). Insights into gathering high-quality material properties is discussed in more detail later in this chapter, but for the meantime, the designers of adaptive ultrasonic transducers incorporating shape memory alloys must contend with practical ranges of elastic or Young's modulus in many cases. It is not unusual for a more specific magnitude of Young's modulus to be provided by a manufacturer, but it will typically be for only one of the transformation temperatures, and commonly these are the A_S or A_F transformation temperatures. The provision of this information should be considered in the contexts of the material fabrication and processing histories applied to the shape memory alloy, and this can be in terms of the chemical composition, the heat treatments applied to the specimens, and the cold working or other machining processes used in the manufacturing phase. Given these fabrication and processing conditions which are inevitable for any transducer component of even marginally complex shape, it makes the elastic properties particularly difficult to determine with precision, and certainly not sufficient for accurate models without several iterations in parallel with an experimental prototyping process in place. Modelling and simulation can be further complicated by the fact that some configurations of ultrasonic transducer incorporate other components which exhibit their own long-standing problems. One example is the epoxy resin bond layers in flextensional transducers, whose physical material properties are also highly sensitive to temperature, and which can often be difficult to quantify. Under vibratory conditions, small cracks and deformations can be generated in epoxy resin bond layers which compromise on the dynamic and mechanical performance of the transducer. A significant challenge is therefore how to quantify the contributions to the dynamic response of the transducer from the epoxy resin bond layers, and how to quantify those from the Nitinol shape memory alloy, to ensure reasonably accurate estimations of Young's modulus for Nitinol. Such problems can be highly complex in the design of adaptive ultrasonic transducers, and the general advice presented in this section is intended to bring the reader's attention to it, with some suggestions for how to manage the challenge. The developments in modelling and simulation, and especially in finite element analysis, are expected to continue to rapidly progress, but the practical application of these simulations depends on high-quality experimental data.

4.2.5 THERMAL ANALYSIS OF PHASE TRANSFORMATIONS

One of the most widely accepted and accurate methods available of measuring the transformation temperatures of Nitinol is a technique developed and patented in the 1960s by Watson and O'Neill (Watson and O'Neill, 1962; Hernández, 2016), called differential scanning calorimetry (DSC). This is a fully established thermoanalytical technique for measuring the thermally dependent physical and chemical properties of a wide range of materials, generally with the major goal of understanding the nature of phase transitions. Specifically, DSC is used to measure the heat necessary to raise a sample's temperature in comparison to a reference (hence the term *differential*). A common output is the measurement of heat flow (the *calorimetry*) as a function of sample temperature, given in terms of the increase or decrease of the temperature at a pre-defined rate (which is the *scanning* part). Such a measurement for a material such as Nitinol allows key phase transition events to be clearly defined from where transformation temperatures between these microstructural phases can be determined with a reasonable degree of accuracy.

Applications for DSC include the analysis of metals which transform between different microstructures within a defined temperature window, such as shape memory materials. Others include the analysis of various polymers, for example, in the identification of their glass transition temperatures (often designated as T_g), or for the analysis of chemicals and their reactive behaviours as a function of temperature. In terms of the DSC process itself, there are two important considerations to be aware of prior to any measurement, expanding on information which can be found in (Feeney, 2014).

The first relates to defining an appropriate standard material for the reference element in the experiment. Reference standard materials are generally those of sufficient purity and stability, where indium or tin are particularly common. It should be evident for scientists and engineers that proper calibration of a scientific instrument is a vital step in the measurement process. This is especially important for accurate transformation temperature data for a shape memory material. In many DSC systems, there can be imbalances in the temperatures or thermal states between a sample and its reference. This can cause a baseline drift, which is a phenomenon directly affecting the accuracy of the DSC technique (Gao et al., 2009). It is therefore vital that calibrations are conducted prior to each individual DSC measurement, although this can be specific to the instrument being used. In general, a DSC instrument would remain in a calibrated state until a new reference or temperature range are selected, or a different rate of change relating to the temperature administered to the sample and reference. This is known as scan rate.

The scan rate is hence the second key consideration for accurate and effective DSC analysis. Depending on the target application and accuracy requirements, a pre-configured or arbitrary scan rate may be sufficient to acquire a suitable data set. However, it should be known that there is a trade-off between scan rate and the accuracy of the DSC measurement. In general, where comparatively faster scan rates are selected, more distinctive and prominent phase transformation events are visible in a DSC thermogram, but this comes at the expense of accuracy. For example, a faster scan rate, whilst allowing a measurement to be completed in a relatively short time and with a clear picture of phase transformation behaviour from the dependency of heat flow in the sample on the temperature, may not yield a sufficiently accurate estimation of the transition temperature to generate a particular material phase microstructure. The reason that heat flow in a DSC measurement rises as a function of scan rate is because the applied temperature change is delivered in a shorter time than it would be for a slower scan rate. The power which is delivered to each furnace housing the sample and reference, respectively, must adjust to this more abrupt change in conditions, and hence there is an associated impact on the power requirement to the system. This brings about the compromise in DSC analysis concerning sensitivity, or the relative prominence of transformation events in a thermogram, and the accuracy of the phase transition temperatures (Shaw, 2008). The advantage of using a faster scan rate is that less prominent phase transition events can be detected in a thermogram. However, by using this faster scan rate, a thermal lag is typically generated in the response, which is directly a consequence of the delay between the DSC system processing the change in conditions to the measured heat flow. It should be noted that it is not unusual for a scan rate of 10°C/min to be applied as a generally accepted scan rate for most applications (Shaw, 2008), but if the time and resources permit, there should be some investigation of these parameters in the thermoanalytical characterisation of the material.

It is worth bringing the attention of the reader to two alternative approaches in the identification of transformation temperature, and although they are not strictly thermoanalytical techniques, may prove to be valuable depending on the resources and time available to the investigator. These approaches are briefly outlined below, where one focuses on the identification of the A_F transformation temperature, and the other makes use of the shape memory effect of a shape memory material. This section expands on the processes as detailed in Feeney (2014).

4.2.5.1   The Active A_F Technique

This is commonly regarded as one of the simplest methods of transformation temperature identification, which does not require particularly specialist equipment, and which can yield relatively rapid results. It can also yield relatively accurate measurement of the final austenitic transformation temperature, A_F, but the principal disadvantages are that it is not possible to obtain any accurate estimation of heat flow, and there will be little confidence in the reliable identification of other transformation temperatures, such as the martensitic transformation temperatures.

This method can also be referred to as the alcohol or water bath test (Feeney, 2014), based on typical approaches applied in the past. As a general overview, a sample of Nitinol to be tested is physically deformed at a temperature below the start temperature of the martensitic phase transformation, M_S, from where it is heated through a pre-defined temperature range. Clearly, if A_F is not known then it would not be expected that M_S would be known either. However, depending on the composition of the alloy under test, the transformation temperatures of Nitinol can be broadly estimated, and this technique is a rapid and practical method of identifying A_F with a generally acceptable level of accuracy. Once the material is passed through the appropriate temperature range, the deformation recovery is usually closely monitored towards full physical recovery of the sample. The temperature at which full recovery is observed can be closely correlated with A_F. This technique is generally applied to specimens of Nitinol wire, although it can be utilised for the analysis of thick sheets or films. In the case of Nitinol wire, it would be feasible for the bend angle in the wire to be monitored as the temperature to which the specimen is exposed, is changed. Here, the bend angle as a function of temperature would yield a clear indication of A_F, and should the start temperature be sufficiently low, A_S, which is the start of the austenitic phase transformation.

Noting previous comments regarding the parameters for this technique to be properly conducted, there will be differences in the start and finish temperatures, depending on the type of shape memory material used and its composition. For example, superelastic Nitinol alloys typically have A_F temperatures much lower than those of shape memory, often in the order of 0–10°C or just below normal ambient room temperature. It has been reported that for superelastic alloys, a viable start temperature for this test would be in the region of −50°C (Feeney, 2014), but it is an approach which needs some careful thought and consideration by the investigator.

4.2.5.2   The Constant Load Method

Whilst the active A_F technique outlined in this chapter is often used for both superelastic and shape memory forms of Nitinol, the constant load method tends to be predominantly utilised in the analysis of shape memory Nitinol (Feeney, 2014). The basis of this method is that a material is subjected to a mechanical load, for example, a stress applied to two ends of a wire, after which point a thermal cycle is applied, from where the transformation temperatures can be estimated by monitoring the physical deformation of the Nitinol specimen as a function of temperature.

This method may be undertaken in the following way. A specimen of shape memory Nitinol (thus, the material is relatively compliant around ambient room temperature, with the stiffer austenite phase estimated to emerge at a significantly higher temperature) is physically deformed in its martensitic condition. An example of this is that it could be bent or twisted out of shape. Note that if the specimen of Nitinol is indeed shape memory, where A_F is significantly higher than ambient room temperature, then there would not normally be a requirement to fix the material in place. A superelastic Nitinol alloy would immediately recover its shape upon the release of the applied load. With the shape memory Nitinol specimen in its deformed state and with a martensitic microstructure, it is then heated until the microstructure is fully austenitic. This can generally be determined by monitoring the physical shape recovery of the specimen to its original state. Finally, the specimen is cooled to below the final martensitic transformation temperature, M_F, where the specimen is once again relatively compliant, but where the original configuration is retained. The proportion of shape recovery can be monitored at all stages of the thermal process, where deformation (e.g., in mm) can be plotted as a function of temperature. It is common to identify a loop, from where the four principal transformation temperatures associated with both the austenitic and martensitic microstructures can be identified. There will be a hysteretic phenomenon which emerges between the data plot for the heating portion of the thermal cycle compared to that of the cooling, in part based on the available energy in the system from the transition to one phase microstructure from another, but this in general does make it somewhat straightforward to clearly identify all four transformation temperatures of interest.

The clear advantage of this method is that it is possible to reliably estimate four transformation temperatures instead of only one, and whilst using relatively simple resources. Methods such as this one, and the active A_F technique, are potentially useful for analysing specimens of Nitinol which are not suitable for traditional mechanical testing approaches, and where obtaining samples of test material is particularly difficult. In general, both the active A_F technique and the constant load method are relatively inexpensive and simple approaches to identifying selected transformation temperatures, especially compared to DSC, with a reasonable level of accuracy. However, since many physical properties of Nitinol are dependent on the phase microstructure and on both temperature and stress, there are other ways in which transformation performance can be measured. For example, the electrical resistivity of Nitinol is dependent on the phase microstructure, and should the test specimen be suited to it, monitoring changes in this property as a function of stress and temperature would allow measures of transformation temperatures. This has already been explored with some success (Antonucci et al., 2007; Brammajyosula et al., 2011), and is worth retaining as a characterisation option. Interestingly, there is some evidence that in using a measure of electrical resistivity for determining the transformation temperatures of different phase microstructures, that this approach is more effective for capturing the transformation temperatures associated with the R-phase than DSC (Antonucci et al., 2007). It should also be noted though that whilst there may be more sensitivity to the existence of the R-phase using this approach, there are also reports of difficulty in data interpretation, and the limitation of obtaining latent or specific heat information (Shaw, 2008).

4.2.5.3   A Typical Thermoanalytical Characterisation of Nitinol

Specific to Nitinol and relevant to the information contained in this book, with extensions to other compositions of shape memory material including shape memory polymers, there are characteristic transition events to watch out for in a DSC analysis. In this brief section, a step-by-step overview of conducting a typical DSC measurement is described, which is intended to be of use to an investigator undertaking such a measurement with a commercially available system. Whilst there may be differences between specific instruments available, the general steps should be consistent. In general, the process can be broken down into the following key stages:

1. Preparation of the sample for testing
2. Calibration of the instrument
3. Data acquisition process
4. Analysis of the data

As a general overview of the first stage in this process, the preparation of the sample for testing, it is usually customary to begin by carefully sectioning a sample of material to install inside the test chamber of whichever DSC instrument is used. Herein lies the first major challenge, as ensuring a high-quality sample of material is vital to ensure accurate thermogram results. Samples of material for typical DSC instruments are in the order of a few millimetres, at most. It is strongly recommended that in the development of any ultrasonic transducer incorporating a shape memory material, sacrificial specimens are obtained where possible, for thermomechanical and thermoanalytical investigations. This avoids having to remove a portion of material better used for integration with an ultrasonic device. In the preparation of a sample, in this case Nitinol, it is advised that a suitably hard cutting blade, for example, diamond, is used to ensure clean and uniform cuts to the sample. Any physical features, such as residual stresses, resulting from plastic deformation of the sample which are generated in the sample preparation process can create distortions in the data from a DSC measurement (Shaw, 2008), and thus should ideally be avoided. There have since been reports into how residual stress can affect the resonant performance of Nitinol devices, for example, through a modelling approach in the work of Bale et al. on how the resonant frequencies of cantilever-type resonators change with different magnitudes of residual stress (Bale et al., 2020). One key observation is the evidence that resonant frequency can increase with residual stress, but much further research needs to be undertaken to understand the links between residual stress, geometrical and material properties of a resonator, and the dynamic properties.

Samples are typically calibrated by utilising a standard reference material which is used as a comparator. Common materials for such purposes include indium and zinc, and they are ideal choices because they can be used repeatably without compromising the quality of the calibration process. It is customary to define certain critical parameters of the standard reference material, for example, the melting temperature and the estimated heat flow. It should be noted that standard reference materials can be specific to a particular instrument, and that indium and zinc are highlighted here as examples from experience.

From experience of working with commercial DSC platforms, ideal weights for certain samples in some DSC tests concerning Nitinol are in the order of milligrams, and so this gives an indication of the quantity of material required to obtain high-quality data. In general, the mass of a sample of material must be measured and configured into the DSC instrument. This is to ensure that accurate heat flow data is determined by the instrument and that clear transformation events occurring in the sample of material with respect to temperature are observed. Once the specimen is ready for testing, it is typically encased within a vented sample pan, where the venting is introduced to ensure there is no damage to the DSC instrument from any excess pressure release from the sample pan when testing.

The second stage of the DSC measurement process is the calibration of the instrument itself. It is evident that the specifics of this stage of the process are unique to whichever instrument is being used, but a few general points can be made, which an investigator may find useful. The first relates to the temperature range. In this case, a calibration can remain valid so long as the defined temperature range of the measurement does not change. This is partly because the baseline temperature data is typically used to subtract from measurement data, thus eliminating noise or other extraneous effects in the system which can bias or influence measurement accuracy. The second point important to highlight at this stage is that of the scan rate. Similar to the case for temperature range, once the scan rate of the system is adjusted, a new calibration should ideally be implemented. What follows in the final stages of an instrument calibration process are typically automated routines relating to the furnace, using information gathered from the preceding calibration steps, and implemented to ensure accurate measurement.

Upon the successful completion of the calibration process for the DSC instrument, test material would typically be loaded into the sample chamber, and a characteristic thermogram produced, similar to that depicted in Figure 4.6 (Feeney and Lucas, 2014), highlighting key transformation temperatures of the material as a function of both heat flow in the sample and temperature.

The key features of note in a DSC thermogram, in the context of assessing a shape memory material like Nitinol for its suitability to an adaptive ultrasonic transducer, are the transformation temperatures, which are directly associated with spikes in heat flow magnitude. The DSC thermogram shown in Figure 4.6 illustrates the temperatures at which martensitic and austenitic Nitinol have been determined. The spikes in the heat flow indicate the transition between these microstructures, and these can be aligned with a change in certain physical properties within the material. For example, in the transition from martensite to austenite, there will be an increase in Young's modulus of the material in the order of several tens of GPa. Magnitudes of properties like density will evidently not

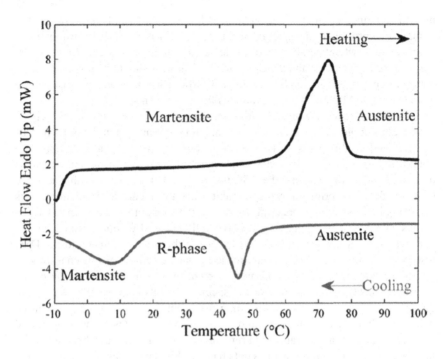

FIGURE 4.6 A typical differential scanning calorimetry thermogram for Nitinol, with indications of the phase microstructures at different zones. Reprinted (adapted) from the work of Feeney and Lucas (2014), under the CC-BY 4.0 licence.

change, but several other properties including electrical resistivity, are highly phase dependent, as described earlier in this book.

The major advantage of being able to use DSC to analyse a sample of shape memory material intended for use in an adaptive ultrasonic transducer is that the DSC instrument can typically accommodate a range of shapes and sizes. The key limitation is that sample size is generally smaller than a few millimetres, but there is some tolerance in this and is generally accounted for in the calibration process. Utmost care should be taken when preparing a sample of material to avoid biasing the results, for example, from residual stresses at cutting sites, but DSC generally provides a rapid and effective way of understanding the transformation events in the material before it is embedded into an adaptive ultrasonic transducer. The sample material used in the DSC analysis would not normally be used, which is why the preparation of sacrificial material should be included in the preparation plans for transducer design.

Once the transformation temperatures are known, as illustrated in Figure 4.6, preparations can be made for how the adaptive ultrasonic transducer fabricated using the assessed material will be operated. However, prior to the design and fabrication of a transducer, it is necessary to provide some instructional information on mechanical testing techniques, with some of their advantages and limitations.

4.2.6 MECHANICAL TESTING TECHNIQUES

The measurement of temperature-dependent phase transformations in the form of transformation temperatures was provided as an overview in Section 4.2.5. This information is critical to enable the control of a Nitinol- (or shape memory alloy) based adaptive ultrasonic device in terms of dynamic response or phase microstructure of the material. The other aspect of the necessary preparatory work in the design and fabrication of adaptive ultrasonic devices incorporating shape memory materials is the application of mechanical testing techniques. This section will briefly provide an overview of common approaches and some of the known limitations.

4.2.6.1 Overview

The primary objectives of conducting mechanical testing of shape memory or superelastic Nitinol, for example, prior to incorporation in an adaptive ultrasonic transducer configuration, are to understand how the measured transformation temperatures align with the physical ability to control the dynamic properties over that designated temperature range, and to determine how the presence of the material in the system will influence the dynamic performance of the device. Mechanical testing methods can provide insights into the mechanical properties of the material, including Young's modulus or damping, and can therefore provide valuable information regarding the proper utilisation and application of the shape memory alloy as it is embedded in an adaptive ultrasonic device.

4.2.6.2 Notable Approaches and Challenges

It has been reported that, in large part due to the continued relatively poor understanding of how the mechanical properties of Nitinol are influenced by the phase transformations generated in the microstructure, the applications for which Nitinol and its alloys are being successfully utilised remains relatively low (Liu and Xiang, 1998; Adharapurapu, 2007). It is known that a range of physical properties of Nitinol, including thermal conductivity and expansion, magnetic susceptibility, and electrical resistivity, will all exhibit changes dependent on the phase microstructure of the Nitinol (Duerig, 2012). This is in addition to the changes generated in Young's modulus of the material for different phase microstructures, already outlined in this chapter thus far. Although some of these properties may exhibit more modest changes than others, for example, the electrical resistivity, the underlying mechanisms are not always evident.

A major concern surrounding Nitinol, observed over several years and in a variety of disciplines, is its fatigue performance. It is known that it is possible to change the mechanical response of Nitinol through forms of cyclic deformation in such a way that fatigue failure has been identified (Kuhn and Jordan, 2002). The significance of this in the context of adaptive ultrasonic devices is that ultrasonic transducers can typically operate in a cyclic manner by their very nature. Therefore, ensuring resilience of the material to cyclic deformations, and a general resistance of the material to fatigue, is important. There may be a certain threshold of applied stress, below which fatigue life can be optimised, or as an alternative, other compositions of alloy can be considered to enhance fatigue life, as has been suggested thus far in this chapter.

Another point of note to raise regarding Nitinol, but not exclusive to its mechanical performance, is that there have been concerns published regarding the presence of nickel in the alloy, and that nickel may be released over a prolonged period of time (Shabalovskaya et al., 2009), which may have significance for medical applications. Future applications of Nitinol are not restricted to the biomedical industry, and it is important to note that the biomedical industry does still make use of Nitinol alloys in a significant way. It is nevertheless important to be aware of any potential risks associated with the material. It is expected that alternative forms of alloy and other compositions of shape memory alloy will come to prominence in the next few years, as the scientific field progresses, and fabrication techniques further develop.

4.2.6.3 Key Techniques

Arguably the most common method of mechanical testing for Nitinol is through tensile testing. The traditional objective of tensile testing is to identify the response of the material to stress–strain conditions, for an applied force in a particular axis and often to assist in the identification of modulus data. A common commercial application of Nitinol is in the biomedical industry, in the fabrication of stents. To achieve this and for certification and regulation, the ASTM F2516 standard will typically be utilised, which is the *standard test method for tension testing of nickel-titanium superelastic materials* (ASTM, 2007; Kumar and Huang, 2023). Crucially, this standard is useful to guide an engineer towards understanding the elastic and plastic properties of the superelastic Nitinol (a common material choice for the biomedical industry), in addition to its strength, under the application of uniaxial-applied tensile loads or stresses. Generally, it is strongly advised that engineers adhere to the relevant standards for investigation. There are several standards available for capturing the properties of interest for Nitinol, or for undertaking robust experimental investigation. For example, and in addition to the above, ASTM F2004-05 is a guide for transformation temperature identification using thermal approaches (ASTM, 2004), and ASTM F2082-15 provides an alternative approach of transformation temperature measurement using a bend/ free recovery method (ASTM, 2016).

Since Nitinol is an inherently hysteretic material, its mechanical response in a loading condition (effectively as tensile loads or stresses are applied) will tend to differ to that for the material in an unloading condition, where the applied stresses are removed. Due to this, the stress–strain profiles which can be produced will be different for a loading case compared to an unloading case. If we consider the cyclic motion of an adaptive ultrasonic device, one can visualise the change in stress–strain profile for each vibrating portion of shape memory material in the device. How the generation of phase transformations in the material in parallel with the applied stresses, and the subsequent impacts on the stress–strain profiles, in part showcases the complexity surrounding the measurement of the mechanical performance of materials such as Nitinol.

A tensile test machine, such as those materials testing systems manufactured by Instron®, is a common choice for the testing and analysis of Nitinol. The equipment available is within the specification required to measure the thermal–mechanical response of many forms of Nitinol, both in terms of geometrical configuration and composition. A standard such as ASTM F2516 for tension testing may be advisable for a sample of Nitinol in wire form, for example, because it is a material configuration conducive to the uniaxial application of stress. This is a challenge for the measurement of mechanical properties of Nitinol when machined or formed into more irregular shapes, such as cymbal transducer endcaps. This factor is critical, because much of the advances that may be seen on a brief background search will refer to the mechanical testing of Nitinol in wire form, and this is understandable given their broad application in the biomedical industry. However, it is not strictly instructive, should the mechanical properties of Nitinol in another configuration, beyond wire, ribbon, or sheet in their as-received forms from the manufacturer. This is because the mechanical and thermal processing histories as applied to the Nitinol can significantly influence the transformational properties and potentially the mechanical performance of the material. The direct consequence of this is that the Nitinol may then behave differently when embedded as a component part of an adaptive ultrasonic device, compared to how it was synthesised to behave. There will also be stresses introduced depending on the chosen fabrication method, whether it be electrical discharge machining or milling with a diamond tip. Such residual stresses may impart changes to the transformational performance of the material, and therefore in terms of its mechanical or dynamic response.

Another factor to consider is that there is strong evidence, specifically in the case of superelastic Nitinol, that the measurement of transformation temperatures using a thermoanalytical technique such as DSC can yield erroneous or inaccurate estimations of transformation temperature (Feeney and Lucas, 2016). This has a direct impact on the correlation of mechanical performance patterns with how the phase microstructure of the Nitinol changes, and so great care must be taken in the analysis of data. A primary cause for such inaccuracies has been attributed to the cold working required to form the Nitinol into the configurations under discussion, and it is hence further evidence that the impacts of mechanical and thermal processing histories must be properly understood, for accurate mechanical property data and device performance to be ensured.

As a further comment on tensile testing, the general point to understand is that it has largely been used to measure plateau stresses and the nature of superelasticity in Nitinol (Favier et al., 2006), and most commonly in wire specimens. There continue to be challenges though, since it is known that discriminating between elastic, plastic, and twinning modes of deformation can be very difficult (Rajagopalan et al., 2005), and this is analogous to understanding which phase microstructure has been generated in a particular heating or cooling cycle, for example, between martensite or R-phase. The difficulties associated with the measurement of elastic or Young's moduli using conventional mechanical testing techniques have been widely reported for several years, including in the 1990s (Liu and Xiang, 1998), where a principal contributor is the generation of alternate phase microstructures. For example, the formation of stress-induced martensite for a specimen of Nitinol above the A_F transformation temperature could potentially become an obfuscation for a measurement (Perry and Labossiere, 2004), if the threshold for stress-induced martensite was not properly understood. Such problems have been encountered, with notable examples in the scientific literature (Liu and Xiang, 1998), and as an alternative, techniques including neutron diffraction have been implemented to derive useful and practical information regarding the measurement of elastic modulus in Nitinol. One example is for the R-phase (Olbricht et al., 2011), but it is also notable that other properties including damping can be highly sensitive to the applied test method (Li and Feng, 1997), and so the choice of characterisation technique is important.

There are several forms of modulus which can be measured for a material, and these can comprise Young's modulus, bulk modulus, and shear modulus as principal properties. There can also be storage and compression moduli to consider too, giving a complex picture of modulus. However, there are differences between static and dynamic cases within all this, and it has been reported that by applying a principle of apparent modulus of elasticity, then this may yield a more representative picture of how a material behaves under certain conditions (Duerig and Pelton, 1994). Modulus is a property which is dependent on several factors, one of which is the processing history of the material, which can include heat treatments such as annealing (Meyer, Jr. and Newnham, 2000). Such heat treatments can be used to eliminate or significantly reduce residual stresses. Nevertheless, it is not unusual to find differing reports of modulus data in the literature (Liu and Xiang, 1998), and this is understandable given the complexity of the material and how its properties can be accurately measured. It is imperative that as much useful information regarding a sample of Nitinol can be gathered as is practically possible. One option is to obtain small samples or cubes of material from a supplier of Nitinol, for thermal–mechanical testing as required in parallel with device design and characterisation. This helps to ensure that the alloy composition matches that of the Nitinol component, but for which there would be no requirement to remove part of that component for the characterisation process. The main limiter is that if one is working with an irregularly shaped Nitinol part, which is highly likely in the case of a component for an ultrasonic transducer, then because its mechanical and transformational properties are highly dependent on its mechanical and thermal processing histories, it is very difficult to obtain accurate measurements of their mechanical and dynamic properties as required. The components are not often in a standard configuration necessary for a standardised measurement method, and there is an argument to be made that by purely undertaking a characterisation of a material, there is a significant risk that its physical response characteristics are changed. Therefore, whilst methods are available to understand the physical features of a sample of Nitinol in whichever configuration it is available, there is some progress still to be made to integrate the characterisation of Nitinol fully and robustly into the transducer design process.

As referred to in Section 4.2.4, one viable approach is to begin with reasonable and informed estimates of modulus magnitudes for each phase microstructure, for example, through the characterisation of sacrificial material obtained from the manufacturer. The modelling and simulation of a transducer in the design phase could then feasibly progress, whereby a practical range of dynamic performance parameters could be defined. Once a prototype device is constructed, the model could be reverse engineered to identify more accurate estimates of modulus magnitude for each phase microstructure. Of course, there is another option for the mechanical and thermal characterisation of Nitinol which has become more popular in recent years, called *dynamic mechanical analysis*.

As an overview, dynamic mechanical analysis is a highly effective and sensitive technique to characterise a range of material properties, including modulus and damping. Commercial instruments available today can do this for multiple degrees of freedom and in different configurations, for example, in tension, for a cantilever case (either single or dual), in compression, and in three-point bending. Some instruments will also be able to apply a shear to a sample. A dynamic mechanical analyser will typically apply loading conditions to a specimen of material, whether it be inorganic or not, and capture the physical properties of interest as a function of frequency, time, stress, atmospheric condition, or temperature. The benefits of the dynamic mechanical analysis process to the study of Nitinol and its performance under both static and dynamic loading conditions with temperature are therefore evident. The main advantage is that the different configurations that commercial instruments can house and accommodate specimens of material mean that one can effectively create a simulated environment for a sample. For example, imagine a Nitinol end-cap subjected to a vibration cycle over a defined temperature range. All that would be required would be a second end-cap to be manufactured alongside the end-cap to be built in to an ultrasonic transducer configuration. It would potentially yield a greater quantity of useful information, beyond that which is possible from a differential scanning calorimeter which will only provide accurate estimates of transformation temperature based on the heat flow in the material. A dynamic mechanical analyser can be operated at sub-zero temperatures via liquid nitrogen cooling, with some instruments capable of approaching temperatures around −200°C, and into the hundreds of °C above zero in a heating cycle. The operational temperature ranges for Nitinol alloys, and many other compositions of shape memory alloy, are therefore within the instrument range, making dynamic mechanical analysis a highly desirable technique in the design and characterisation process for an adaptive ultrasonic device incorporating a shape memory alloy.

Of course, no characterisation method is without its limitations. One is that there have been reported discrepancies between the transformation temperatures measured using both electrical resistance monitoring and differential scanning calorimetry, compared to those identified using dynamic mechanical analysis (Vilar et al., 2023). One explanation which has been given is that there is a notable level of heat transfer present between the clamping fixtures and the sample, and so there should be some caution exercised when using a single mode of material characterisation. It is advisable to consult a variety of experimental techniques, where possible, to capture a broad picture of the thermal and mechanical responses of Nitinol, prior to their incorporation in an adaptive ultrasonic device. Nevertheless, the potential of dynamic mechanical analysis is significant for the trial and validation of different forms of shape memory material for integration with adaptive ultrasonic transducers.

4.2.7 PRACTICAL ADVICE FOR DEVICE INTEGRATION

The most widely known and reported example of integrating shape memory materials into ultrasonic transducers, to generate adaptive features, is the cymbal transducer. Fundamental research in this area has yielded key lessons for future incorporation of shape memory materials into ultrasonic devices, which will eventually encompass a broad range of materials, for example, those referred to in the literature, such as Jani et al. (2014) and Alipour et al. (2022). This section will focus on an overview of the design strategy developed for an adaptive cymbal flextensional transducer, which can be applied to other configurations of ultrasonic transducer. The intention here is to summarise several important recommendations for transducer design utilising the shape memory material Nitinol. The design and fabrication of an adaptive ultrasonic transducer (e.g., the cymbal) must consider, as a guideline minimum, the following factors:

1. Material or alloy selection
2. Measurement of transformation temperatures
3. Transducer assembly technique
4. Characterisation protocols
   - Temperature measurement
   - Mechanical response characteristics
5. Modelling and simulation

It has been established that the mechanical behaviour of the shape memory alloy Nitinol is complex, principally in how its elastic properties change as a function of both temperature and stress. This has significant implications for the design of an adaptive ultrasonic transducer from first principles. A 2024 paper has demonstrated the real difficulties in the measurement of elastic modulus in Nitinol (Liu et al., 2024b), where among others, factors such as the orientation of the material under study and its physical condition directly influence the magnitude of the moduli. In such studies, techniques including resonant ultrasound spectroscopy are used, where it is evident that great care should be taken in considering the validity or usefulness of the measured properties. It should not be assumed that the elastic modulus of the material measured using a particular technique like resonant ultrasound spectroscopy will directly translate to the elastic modulus of the shape memory material when integrated into an ultrasonic transducer configuration. Factors which influence the material characteristics include the mechanical condition of the material in response to its hot and cold working histories, the boundary conditions on the materials embedded in the transducer configuration, the conditions of the environment in which the material is operating, and the composition of the shape memory alloy.

First, and considering the practical limitations associated with Nitinol and other shape memory materials, the most effective place to begin the engineering of an adaptive ultrasonic transducer is to consider suitable alloy selection, appropriate to the intended application. Taking Nitinol as a feasible example, it is sensible to utilise materials which are either widely available or relatively straightforward to synthesise. There will be cost implications for utilising shape memory materials which are difficult to formulate, irrespective of their physical advantages in terms of shifts in elastic modulus or other material properties. Adaptive ultrasonic transducers for practical industrial and medical applications must be cost effective, in addition to exhibiting their intended transformational and adaptive dynamic responses. Therefore, before further fundamental research is progressed, there may be limitations on the viability of more complex or intricate compositions of shape memory material for integration with ultrasonic devices for commercial industrial or medical applications. The discussion from this point forward is limited to Nitinol, where it is assumed that the reader can take forward some of the lessons and apply to other compositions of shape memory material with adjustable elastic properties, for example, those which are dependent on temperature.

It is now important to consider the loading conditions under which the adaptive ultrasonic transducer will be in operation. It has already been established that phase transformations in Nitinol are dependent on stress, at least in part, and so it can be reasonably assumed that beyond a particular stress limit (arbitrary at this stage), then the material, and hence the transducer, will need to be subjected to a higher temperature than the case of an absence of any mechanical or dynamic loading. The relationship between stress and temperature for a shape memory alloy such as Nitinol has been widely reported (Kumar and Lagoudas, 2008), but in general there is a linear relationship between the applied stress and the temperature to which the material is exposed, which has direct implications for the temperature required to reorient the microstructure of Nitinol between its austenitic or martensitic (twinned or detwinned) states. It should be noted that through such microstructural reorientation, sufficient stress at a temperature above that which generates the final austenitic microstructure, will reorient the microstructure to martensite. As shown previously in this chapter, this is the nature of the superelastic effect. For this purpose, it is not unreasonable to consider superelasticity in the mechanical and dynamic responses of Nitinol integrated with an ultrasonic device, particularly one designed for a power ultrasonic application. The major challenge is identifying where that transition point is, and how such behaviours will manifest in the dynamic response of the transducer. These material challenges are significant, but there is added complexity because the physical characteristics of Nitinol, whilst universal, are also in many ways unique to the specific configuration of transducer designed. A particular thickness of Nitinol cap, or variation in end-effector shape, are all realised through adjustments in hot and cold working processes used to fabricate these materials. From this, there would hence be differences in material response from one configuration of transducer to another, further complicated by the fact that dynamic nonlinearities inherent in ultrasonic devices are also typically unique to different configurations of ultrasonic transducer. These responses may all follow a particular form, but their prediction and measurement are challenging and often individual to a given configuration of transducer. Another important consideration for a material like Nitinol is that its stress–strain response is inherently nonlinear, exhibiting a hysteretic behaviour where the stress magnitudes as a function of temperature are different for a loading condition compared to unloading. The reason for this is that there is a net dissipation of energy in a transformation cycle, which is also dependent on the material and operational conditions of the material. Given a phenomenon like hysteresis, it also therefore cannot be assumed that in a single cycle of mechanical vibration experienced by a transducer component fabricated from Nitinol, that there will be a linear or reversible loading and unloading process. The loading phase of a single vibration cycle may constitute a different amplitude or stress level in the material, at least locally, in comparison to the subsequent unloading cycle.

For ultrasonic transducers in general, loading conditions can be first established through the permissible electrical excitation conditions applied to the piezoelectric ceramics in the transducer assembly. It is customary to then mathematically model, or simulate through finite element methods, the proposed transducer response in terms of its electrical properties and dynamic behaviour. These approaches can yield useful information regarding the likely displacement amplitude and stress behaviours in the transducer and provide indicators of potential failure modes in the transducer materials. The limitations of finite element methods have been addressed in Chapter 3, but one important characteristic which can be challenging to determine and account for in simulation is the damping parameters, such as Rayleigh damping. Such forms of damping are directly linked to mass and stiffness of a mechanical structure, and they are difficult to experimentally verify for a simulation for an accurate determination of the amplitude and stress conditions of a transducer. The use of a finite element simulation to assist in the design of an adaptive ultrasonic transducer based on Nitinol, whilst a key step in the development process, should be considered with care.

As shown already in this chapter, obtaining accurate elastic moduli magnitudes for given environmental and mechanical loading conditions is very difficult, and it may be more practical to engineer an adaptive Nitinol transducer where the experimental process is undertaken in parallel with simulation. It is not difficult to understand why many manufacturers of commercially available Nitinol tend to either avoid specification of the elastic moduli of their material, or define a range of magnitudes, typically in terms of GPa, in which the elastic moduli of their material typically lie. In any case, selection of a suitable alloy of Nitinol can only realistically be made using the available information from a commercial manufacturer, unless there are the facilities and capabilities available to synthesise material bespoke to the application or device of interest. Commercial manufacturers will commonly provide a selection of information to aid in the acquisition of a suitable material for an adaptive ultrasonic transducer. A concise summary of this information is provided in the list below.

- **Active final austenitic transformation temperature**, although on occasions other transformation temperatures are provided, including that related to the start of the phase transformation to austenite, and thus should be clarified upon procurement.
- **Elastic moduli ranges**, for both the material in austenitic and martensitic microstructures. Often this information is not provided, but if it is, then it may be approximate.
- **Information regarding purities**, including carbides and intermetallic oxides, which is important information particularly for biomedical applications.
- **Details of additions and alloying elements**, for example, Cr for enhanced stiffness, Cu for increased transformation temperatures or reduced hysteresis in the mechanical response, or Nb for wider hysteresis.
- **Density**
- **A statement of compliance with relevant standards**, such as ASTM F2063-18 (*Standard Specification for Wrought Nickel-Titanium Shape Memory Alloys for Medical Devices and Surgical Implants*) for biomedical stents (ASTM, 2018).

This list is not exhaustive, but it is a general and realistic overview of what one may expect as baseline information from a supplier. As a comment on the transformation temperatures, it is also customary, though not in each case, to only be given the final austenitic transformation temperature. This is because many industrial and medical applications only rely on this temperature and tends to be the only one of importance for specific mechanical and dynamic systems. For example, one-time use Nitinol-based stents only need to change shape and expand once, above this final austenitic transformation temperature, and so the other austenitic and martensitic transformation temperatures are of little relative importance.

The information available in the material selection stage can be viewed from the perspective of the final austenitic transformation temperature, A_F. This value tends to be expressed with an available tolerance, for example, in the order of $\pm$ 5–10°C, but its real importance is judging which available alloy should be selected for the application of interest. In general, two types of Nitinol alloy can principally be selected. The first is shape memory Nitinol, where A_F exists above what we would reasonably regard as ambient room temperature. This means that the Nitinol can be physically deformed or manipulated at any temperature below this A_F threshold, but once the material has been heated past A_F, the original set material shape will be recovered. This is the shape memory effect, arising from the dislocation motion between the martensitic and austenitic phase microstructures.

It should also be noted that such a transition between phase microstructures does not require physical deformation of the Nitinol, where it will manifest as a change in the material's elastic modulus regardless. This is the fundamental basis for the operation of adaptive ultrasonic transducers fabricated using shape memory alloys. The second class of Nitinol is superelastic, where A_F is this time either around the conventional definition of ambient room temperature or below it. This chapter has already discussed how superelastic Nitinol is most popular for bio-medical applications, such as part of stents, but it has been given little attention in the design of ultrasonic transducers. It is characterised by relatively high strain recovery performance in response to applied stress, where a reorientation of the phase microstructure (from austenitic to martensitic) occurs when the material is at a temperature above A_F. There are hence interesting implications of this in terms of ultrasonic devices and their operation, such as frequency agility for ultrasonic devices operating above a specific pre-defined amplitude threshold, but this has yet to properly be investigated. In terms of the adaptive cymbal transducer fabri-cated using Nitinol end-caps, the level of stress required to generate phase changes in the material has not been studied in detail.

We are inevitably limited in many ways by intellectual property restrictions relating to materials and devices, and so it is important to be well equipped where possible to understand how to measure key properties of these materials in the design, fabrication, and operation of adaptive ultrasonic devices. Most Nitinol alloys have a practical operational limit of 100°C, above which transformational properties are not observable or practical for application. Also importantly, and outlined in some detail in Chapter 3, temperature rises in piezoelectric materials are known to result in nonlinear behaviours in ultrasonic devices, and so this rein-forces the need to properly account for both temperature and stress in the design of an adaptive ultrasonic transducer fabricated from Nitinol. It is useful that com-mercial manufacturers may undertake a thermoanalytical measurement of the material prior to its distribution, where the manufacturing processes are well con-trolled to a sufficient level to ensure a high degree of consistency across different batches of stock. However, it is optimal in terms of engineering practice to con-firm any commercially stated transformation temperatures using available ther-moanalytical techniques. Such measurements of transformation temperatures can also yield information not available from commercial manufacturers, including a more comprehensive understanding of transformation temperatures, and how composition of the alloy, the hot and cold working from the machining processes, and any other processing conditions applied to the material have influenced these transformation temperatures. Nitinol is notoriously difficult to machine (Hodgson and Russell, 2000; Liu et al., 2024b), usually requiring a process like electrical discharge machining to generate the required precision cutting for high-quality components. It is vital that processing of the material is thereby accounted for in the design and fabrication of an adaptive ultrasonic device using shape mem-ory alloys.

An overview of how transformation temperatures through thermograms can be obtained using a DSC instrument has been provided in Section 4.2.5. This approach is highly effective, but it is a measurement technique and does not provide any mechanism for influencing transducer design. Using combinations of heat treatment and changes to alloy composition, the temperatures at which shape memory materials exhibit their phase change can be controlled (Liu et al., 2024b). This is a key factor in the early reports and developments of adaptive ultrasonic transducers which is postulated to become more important over the coming years. There will be concerted efforts towards device controllability, especially within acceptable boundaries of temperature, with further focus on the response stability of a device, for example, how repeatable the switch from martensitic Nitinol is to austenitic Nitinol, and vice versa.

Moving on from the selection of suitable transducer materials and the characterisation of their key properties for adaptive ultrasonics, a fabrication and assembly protocol must be defined for the transducer configuration of interest. Up to this point, there have been several references to a selection of transducer configurations, from the Class IV and V flextensional transducers to the bolted Langevin, and the flexural. It is acknowledged that the Langevin transducer is arguably the most common configuration for innovators in power ultrasonics, but purely in terms of volume, there is a case to be made that the number of flexural ultrasonic transducers available commercially in the automotive industry alone is in the millions. Whichever transducer configuration is of interest to a designer, and in whatever quantity, the fabrication and machining of the shape memory material into the required physical size and shape will likely be a time-consuming part of the transducer fabrication process. Nitinol, and other shape memory alloys, can take a relatively long time to properly process and machine into the desired shape, alongside the underlying design.

In terms of manufacture, electrical discharge machining is a viable option for the machining of extremely hard, stiff, or materials that are generally difficult to work with like Nitinol. It is often referred to as spark erosion, because sparks or discharges are used to remove material from a sample through electrodes. A key aspect of the electrical discharge machining process is that there is no contact between the machining tool and the target material. One type of electrical discharge machining which has found success for Nitinol is the sinker category, where the tool and target material to be machined are submerged in an insulating fluid like oil, where an electrical potential is established between the two bodies. Given the physical nature of Nitinol and alloys like it, there is virtually no other viable machining method available at present, except using blades with diamond tips. The advantage of electrical discharge machining is that it is possible to introduce intricate features with a high level of precision, but the major drawbacks are that it can be a costly and time-consuming process, and that there are still limitations on the geometries it is possible to machine.

A future manufacturing route for adaptive ultrasonic transducers incorporating alloys which are difficult to machine and process using conventional methods, like Nitinol, is through additive manufacturing. There are numerous reports of successes in the additive manufacturing of Nitinol alloys in recent years (Walker et al., 2014; Zhu et al., 2021); however, a major uncertainty in the field of ultrasonics is that the mechanical performance and fatigue life of such additively manufactured specimens are not yet fully understood. Furthermore, there is likely a fine balance of the material properties and additive manufacturing parameters required to properly tailor the alloy for fabrication. The propagation of ultrasound using transducers incorporating additively manufactured shape memory materials also requires significant investigation.

If we take a closer look at the key transducer configurations considered in this book, it is evident that the most significant challenge for the fabrication and assembly of an adaptive ultrasonic transducer relates to the Langevin. A flexural ultrasonic transducer composed of Nitinol would only require a flat Nitinol disc as a minimum to serve as the vibrating plate or membrane, a configuration not overly difficult to engineer using stock material from a commercial supplier, for example. There are numerous reports of the cymbal transducer incorporating Nitinol end-caps in the scientific literature (Meyer, Jr. and Newnham, 2000; Feeney and Lucas, 2014, 2016, 2018; Smith et al., 2022), where end-caps were fabricated using approaches including machining from sheet material, or punched from sheet samples. Similarly, the class IV flextensional transducer configuration incorporates end-caps that can be machined from a specimen of Nitinol sheet, or alternatively, shape-set using heat treatments from a flat rectangular sheet specimen. However, the Langevin transducer generally requires pre-stressing of the entire stack, where the end-masses and the piezoelectric materials are compressed together in the interests of boosting resilience to elevated levels of stress in operation (particularly relevant to the piezoelectric ceramic rings) and to optimise the energy transfer through the device. To facilitate this pre-loading process, it is customary to introduce a machined thread into one of the end-masses, in this case which would ideally be fabricated from Nitinol. This evidently presents a significant challenge for the fabrication and assembly of a Nitinol Langevin transducer, not least because of the intricacy and low tolerance associated with the geometrical features of a thread. Sinker electrical discharge machining is an ideal process for this, and early achievements in this area have been made (Liu et al., 2024b). One factor to consider in the assembly of the Nitinol Langevin transducer, where the end-masses are composed of Nitinol, is the influence of pre-stressing the stack on the transformation temperatures of Nitinol and the subsequent impacts on the dynamic performance of the device. These factors are very difficult to predict in the design process, but they should be considered in how the device is implemented for practical applications.

There are of course other ways in which Nitinol can be embedded into a Langevin transducer configuration. This is not limited to the end-masses, where, for example, the cascaded configuration of transducer has received attention in the literature (Li et al., 2021). Beyond this, an adaptive ultrasonic transducer in the form of a cascaded transducer fabricated with a Nitinol ring in the middle has been used to demonstrate the potential for ensuring stability of resonance across a broad temperature range (Liu et al., 2024b). This transducer is depicted in Figure 4.7, alongside an example of its characteristic response. The interesting benefit of this is that it would be feasible to operate a Langevin transducer at

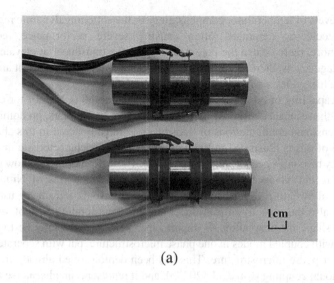

(a)

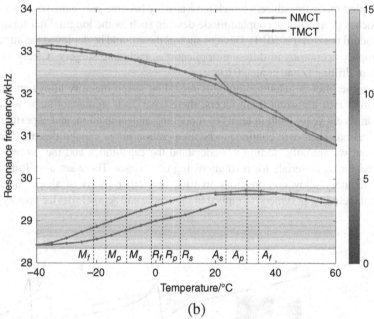

(b)

FIGURE 4.7 (a) The Langevin ultrasonic transducer in a cascaded form, with a Nitinol central ring (NMCT, top) and a titanium central ring (TMCT, bottom); (b) the resonance stability of the NMCT as a function of temperature, compared to that of the TMCT. Note that the colour bar on the right shows the half-power bandwidths. Reprinted (adapted) from the work of Liu et al. (2024b), under the CC-BY 4.0 licence.

higher levels of temperature, compensating for the softening dynamic response of the piezoelectric ceramics, often causing severe performance reductions. Furthermore, there is also the potential to drive the transducer at elevated excitation voltage levels, thereby ensuring higher voltages and displacement amplitudes can be achieved for a single transducer configuration.

The capability of an ultrasonic transducer incorporating Nitinol for control in terms of the resonance response has been noted for several years, predominantly in the flextensional configurations of transducer, as already noted in this chapter. An example of this is illustrated in Figure 4.8 alongside its characteristic impedance-frequency response (Feeney and Lucas, 2014). These developments show two notable and recent advances in the integration of different forms of Nitinol shape memory alloy into different configurations of ultrasonic transducer used today. There is also the capacity of transducer designs incorporating Nitinol, or another form of shape memory alloy, to impart other benefits. One may be to engineer devices with coupled modes in one phase microstructure, but with separated modes in another phase microstructure. This has been demonstrated already in terms of active modal coupling (Liu et al., 2023b), and it represents an alternative approach for coupled mode transducers which have already become prevalent, such as those approaches proposed for coupled mode devices such as the longitudinal-torsional (Al-Budairi et al., 2011, 2013). Devices such as these could be deployed for intricate or complex operations in different applications, including surgery (Cleary et al., 2022), or drilling (Zhao et al., 2020).

As some brief concluding thoughts regarding shape memory alloys and their integration into ultrasonic transducers, there have been significant steps forward made in recent years, both in terms of modelling and simulation, and experimental investigation. There will continue to be acceleration in our capabilities in all these areas, but we are only starting to understand the capabilities and the potential of these advanced materials for revolutionising ultrasonics. There are a multitude of shape memory alloy compositions to trial, and other forms of shape memory material including polymers, to embed in ultrasonic and acoustic device prototypes.

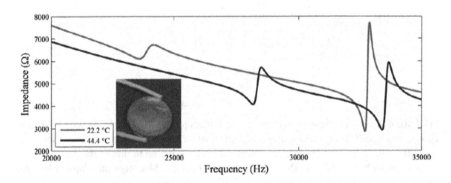

FIGURE 4.8 A Nitinol cymbal transducer (inset) and its characteristic response at two different temperatures, showing control of resonance. Reprinted (adapted) from the work of Feeney and Lucas (2014), under the CC-BY 4.0 licence.

It will require continued efforts from materials scientists, engineers, and technologists to drive this change, and the uptake from industry to translate the successes from these developments into practical applications.

4.3 METAMATERIALS

4.3.1 OVERVIEW

The second major class of advanced material which can be considered in the design and manufacture of adaptive ultrasonic devices is the metamaterial. As a general overview, a metamaterial is that which exhibits properties not usually encountered in nature or in natural materials, and typically requires some level of engineering to realise. In principle, a metamaterial is designed to manipulate a physical phenomenon in such a way that gives some level of control to a process. For example, an optical metamaterial may be designed and synthesised to generate a negative refractive index (Cummer et al., 2016). This type of metamaterial is often referred to as a negative-index material (NIM) and has applications in communications (Triki and Kruglov, 2022), such as how electromagnetic signals are transmitted or received. In general, a metamaterial commonly comprises a series of characteristic design considerations, including material properties, which directly influence its performance. These can include the inclusion of repeated features with defined shapes and sizes that together constitute a larger singular metamaterial structure, the definition or control of the global size of this metamaterial structure, the materials used in the fabrication process, and how the metamaterial is integrated into the target structure or environment.

As a field of research, metamaterials science has rapidly expanded and accelerated with the advent of advanced fabrication capabilities such as additive manufacturing. There are now possibilities across different types of material, whether they be metallic, plastic, or composite. Furthermore, the speed, scale, and precision of available manufacturing techniques has resulted in a significant shift towards advanced resonators not previously possible using conventional fabrication methods. The aim of this part of the book is to outline a selection of key advances in metamaterials, with a particular focus of the discussion on acoustic metamaterials. The field of metamaterials (across the sub-disciplines of design, modelling, manufacture, characterisation, and application) is now significantly broad, and there is evidence that the opportunities afforded from many material concepts previously limited to mathematical modelling and simulation are now becoming reality. For this reason, highlights of some key advances in the development of acoustic metamaterials follows for the remainder of this chapter, as relevant to adaptive ultrasonic devices.

4.3.2 ACOUSTIC METAMATERIALS

The acoustic metamaterial is sometimes referred to as a phononic crystal, and it can be compared as an analogue with the photonic equivalent in optics. An acoustic metamaterial can be considered as a structure assembled of a series of subwave-length resonators, more commonly referred to as meta-atoms (Zhang et al., 2023). These meta-atoms are in principle smaller than the wavelength being radiated. Acoustic metamaterials are entirely artificial, and they are synthesised with preci-sion to enable the production of key features, predominantly dynamic in nature. The general purpose of an acoustic metamaterial is to enable the manipulation or control of sound, where this can be achieved to shape sound wave profiles, cre-ate acoustic bandgaps where sound cannot travel, or to enable amplification at frequencies of interest (Cummer et al., 2016; Gardiner et al., 2021). In principle, and with available manufacturing techniques alongside mathematical models, the field of acoustic metamaterials presents a significantly wide opportunity for the control, modification, and adaptivity of ultrasound waves, and the dynamic response of ultrasonic devices composed of acoustic metamaterials.

Developments in metamaterials throughout the 20th century directly led to the conception and formulation of the acoustic metamaterial. The general understand-ing is that the relevant form of metamaterial with importance for acoustic or sound wave applications was originated by Victor Veselago in the 1960s (Flores-Méndez et al., 2020), who demonstrated the principle of negative refraction. These prin-ciples were eventually put into practice around 2000 by Sir John Pendry based on significant foundational research, from which time the research and development of acoustic metamaterials and their applications significantly accelerated. Relatively early work in acoustic metamaterial design and fabrication consisted of the split-ring resonator (RoyChoudhury et al., 2016), which led to the production of negative refractive index materials, and the development of lenses consisting of acoustic metamaterials. As an example of what an acoustic metamaterial may look like in practice, the schematics provided in Figure 4.9, extracted with permis-sion from the research of Zeng et al. in 2013 from their work on the development of acoustic metamaterials, exhibits both negative bulk modulus and mass density (Zeng et al., 2013).

The correct definition of the spherical and tubular components of the acoustic metamaterial are important, because electromechanically they were demonstrated to enable the energy storage and dissipation required for the distribution of the sound field in the material to achieve the desired material properties. This is where the capacitance and inductance analogues are useful, as illustrated in Figure 4.9. The metamaterial structure shown was developed to exhibit negative mass density and bulk modulus, and negative refractive index was also found (Zeng et al., 2013). Conventional materials are used in the fabrication, including steel, and the research shows how some relatively straightforward manufacturing combined with precise underlying electromechanical modelling can be used to define a suit-able acoustic metamaterial for applications that can include acoustic cloaking (Zeng et al., 2013). This is a structure or mechanism permitting sound waves to pass in a way which renders the target structure effectively invisible to sound. This research also briefly summarises a suitable design approach to using metamateri-als, where it would be possible to design a transducer incorporating an acoustic metamaterial to control acoustic waves through a structure.

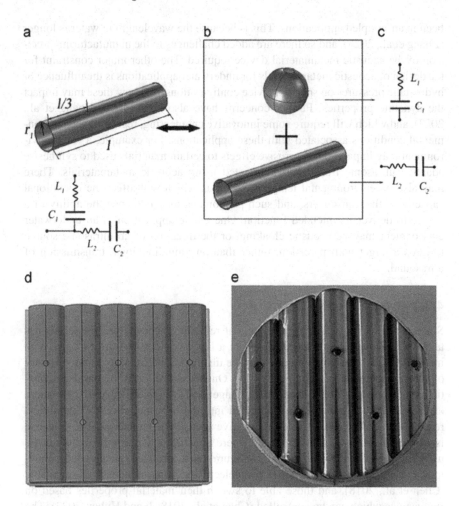

FIGURE 4.9 The tubular-type acoustic metamaterial developed by Zeng et al. in 2013, showing a schematic of the metamaterial (a), the composite meta-atom structure used to construct the metamaterial (b), the effective circuit of the structure (c), and the schematic of the acoustic metamaterial (d) and the fabricated structure (e). Reprinted from Solid State Communications, Vol. 173, Zeng et al., Flute-model acoustic metamaterials with simultaneously negative bulk modulus and mass density Page 15, Copyright (2013), with permission from Elsevier.

In a similar way regarding the developments in shape memory materials and the current status of their integration with ultrasonic devices, the field remains in a relatively developmental state, and there continues to be innovation in medical and industrial applications to qualify acoustic metamaterial concepts. For example, research has grown in the field of underwater acoustics, incorporating acoustic metamaterials into liquid environments with the likely eventual practical adoption in sonar technology. Much of the innovation thus far in acoustic metamaterials has been in air-coupled applications. This is because the wavelength in water is longer (Zhang et al., 2023), and so there are added challenges to the manufacturing precision of the acoustic metamaterial device required. The other major constraint for the design of acoustic metamaterials for underwater applications is the influence of hydrostatic pressures on specific device configurations, and how these may impact the dynamic properties. Further concerns have also been raised (Zhang et al., 2023), and which will require some innovative engineering, regarding the environmental conditions associated with these applications. For example, a water environment may impart some corrosive effects to certain materials used to synthesise adaptive ultrasonic transducers fabricated using acoustic metamaterials. There may also be environmental temperature limits which will affect the operational capacity of the transducers, and such phenomena may influence the ability of a device to deliver its intended function. One viable application in an underwater environment may be acoustic cloaking, or the direction of an ultrasonic source towards a target with precision, rather than an omnidirectional transmission of ultrasound.

4.3.3 Modelling and Design

Since much of the practical development regarding acoustic metamaterials has taken place since around 2000, there are a few key steps forward to those with interests in adaptive ultrasonics. First, the distinction between two major classes of acoustic metamaterial can be clarified. One classification is the **passive** acoustic metamaterial (Ji and Huber, 2022; Zangeneh-Nejad and Fleury, 2019), where external energy is generally not directed into the acoustic waves. A membrane resonator would be an example of a passive acoustic metamaterial. The converse is an **active** acoustic metamaterial, where specific active components can be used in an acoustic metamaterial configuration to produce an intended behaviour. For example, there are active piezoelectric acoustic metamaterials available (Chen et al., 2018), and those able to switch their material properties based on frequencies which can be controlled (Chen et al., 2018; Ji and Huber, 2022). The scope of possible active acoustic metamaterials is significant, and there are many opportunities for tuneable (e.g., in terms of operational frequency band) structures to benefit a wide range of industrial and medical applications. Achieving wider bands of frequency is critical to delivering a new generation of devices based on acoustic metamaterials for cloaking, sound reduction or suppression, the focusing of ultrasound, and enhancements in the transmission of ultrasound or imaging. In general, a clear opportunity for adaptive ultrasonic devices is contained in the subcategory of active acoustic metamaterials, and there are a wide range of additional behaviours which can feasibly be produced such as responses to changing levels of temperature or pressure.

Active acoustic metamaterials constitute a popular area of research, and the pervasiveness of piezoelectric materials and their wide and important scope of application is arguably a major contributor to this. The discussion from this point forward will focus more predominantly on active acoustic metamaterials, because of their general relevance to adaptive ultrasonic devices and the fact that piezoelectric materials are so widespread in current applications. A principal active acoustic metamaterial configuration incorporating piezoelectric materials is the generation of resonant bandgaps in the vicinity of electrical or mechanical resonance (Ji and Huber, 2022). The key advantage of this is that an element of device tuneability can be readily introduced, for example, effective stiffness or mass, and commonly related to Kirchhoff–Love plate theory. We are likely only at the start of an extensive pathway towards highly advanced and tuneable resonant active acoustic metamaterial devices incorporating piezoelectric materials. However, key applications for such devices will be the control and shaping of ultrasound wave propagation fields and the control and adaptability of various electromechanical properties of a device. A major current limitation, which innovations in materials science will continue to address, is how available compositions and configurations of piezoelectric material can be practically incorporated into a device based on an active tuneable acoustic metamaterial. Many highly effective piezoelectric materials with the major electromechanical properties well suited to ultrasonic applications are of the bulk variety, but these can be mechanically brittle (Ji and Huber, 2022), and they can hence place an undesirable constraint on device design.

A key relationship of interest regarding acoustic metamaterials, and important for the consideration of how adaptive ultrasonic devices may be designed in future using these materials, is Equation (4.2) which shows the governing relationship for the propagation of acoustic waves in a homogeneous medium (Ji and Huber, 2022).

$$\nabla^2 P - \frac{\rho}{\kappa}\frac{\partial^2 P}{\partial t^2} = 0 \qquad (4.2)$$

Here, P is the sound pressure and the mass density and bulk modulus of the medium, respectively, are ρ and κ. In this equation, there is an assumption that there are no additional sound sources present in the internal structure of the material. It is customary to also define the specific acoustic impedance Z of the material, which is the pressure divided by the velocity (Ji and Huber, 2022), or alternatively as Equation (4.3).

$$Z = \sqrt{\rho\kappa} \qquad (4.3)$$

It is possible to administer design protocols to active acoustic metamaterials which are typical of other configurations of resonant or dynamic device. There are finite element approaches reported in the literature, for example, (Akl and Baz, 2012), (Allam et al., 2016), and (Petrover and Baz, 2020), and mathematical analogues can be useful in generating effective physical parameters by which device configurations can be formulated. In general, ρ and κ are the key parameters of interest for acoustic metamaterials (Liu et al., 2020), and it is these properties which are the focus for engineering acoustic metamaterials with negative effective, or equivalent, modulus and density. This is generally achieved by integrating units of local resonance into the material, such that the relationship between the acoustic wave field and the material is optimised. As an illustrative example, Huang et al. demonstrated a mass-spring structure with a series of local resonators, to achieve an equivalent negative elastic modulus of the material (Huang and Sun, 2011). The important aspect of this theoretical innovation is that the effective modulus is dependent on frequency, where Young's modulus can be determined to be negative in a generated bandgap. If the effective modulus and displacement ratio is calculated as a function of the normalised frequency, then the zones of negative effective modulus can be determined. A concept design for this has been proposed by Huang et al. and is shown with permission in Figure 4.10 (Huang and Sun, 2011).

In this proposed solution, a base structure would be fabricated, onto which springs would be attached with relatively heavy or concentrated masses placed at the apexes. These masses then become local resonators, with their associated effective masses, damping, and stiffness coefficients. A material of this nature could theoretically be used in a range of applications related to the control of acoustic or ultrasonic wave propagation, or as a filtering mechanism for undesirable waves in a system. It is important to clarify that this advance in modelling and simulation was proposed in the 2010s, but even since then, there is still significant progress to be made in terms of manufacturing and the practical application of metamaterials in acoustics. A selection of manufacturing techniques which can be adopted in the fabrication of acoustic metamaterials, and a detailed overview of the associated challenges, is explored in the forthcoming section.

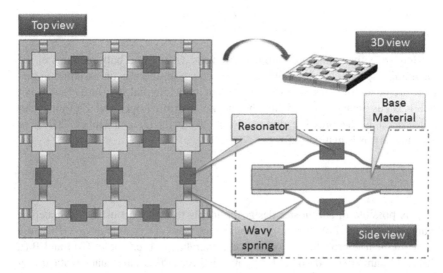

FIGURE 4.10 A proposed solution for realising an acoustic metamaterial with tuneable bulk modulus. Reprinted from the Journal of the Mechanics and Physics of Solids, Vol. 59, Huang and Sun, Theoretical investigation of the behavior of an acoustic metamaterial with extreme Young's modulus, Page 2076, Copyright (2011), with permission from Elsevier.

Before we get to that stage, we can consider another approach to the modelling of acoustic metamaterials. Instead of defining a series of functional units with tailored mechanical properties, as the brief example above has shown, the opportunities afforded by the *Helmholtz resonator* can be outlined. There are a few other possibilities for acoustic metamaterials, including thin films which can be used to suppress noise at certain frequencies, and coiled configurations to expand the operational frequency range and to boost manufacturability (Liu et al., 2020), which may be explored. However, an advantage of an approach utilising Helmholtz resonators is that it offers the possibility to engineer devices for lower frequency design without requiring significant additional mass, because they make use of the resonant characteristics of a fluid such as air. This means that for any prospective frequency range of interest, they can theoretically be manufactured to be relatively lightweight compared to some other configurations of metamaterial.

A Helmholtz resonator exploits the Helmholtz resonance of air bounded in a cavity volume. This is effectively a container, very much like the schematic shown in Figure 4.11, extracted from the work of Dogra and Gupta (2021), which has a narrow aperture at one end that is fundamental to the structure's ability to filter sound waves at specific frequencies, depending on the structural configuration of the Helmholtz cell. Extending the concept of negative elastic modulus, it has already been proposed that the Helmholtz cell can be used to realise a structural configuration of metamaterial with a negative effective elastic modulus (Liu et al., 2020). This was conceived by demonstrating that a transmission bandgap can be created if the frequency of the sound wave is opposite to that of the frequency at the exit of the Helmholtz resonator cavity. Therefore, modelling and design of

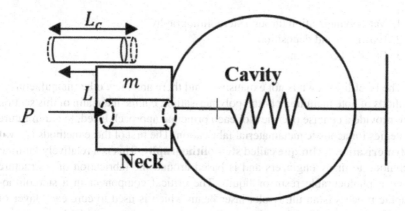

FIGURE 4.11 The Helmholtz cavity showing an example of a unit cell as a resonator, where a series of these could be used in an acoustic metamaterial. Here, P denotes the sound pressure of an approaching acoustic wave, m is the mass, here relating to the air in the neck part of the resonator, L is the length of the resonator neck, and L_c indicates the effective neck length to allow a correction factor to be applied. Reprinted (adapted) from the work of Dogra and Gupta (2021), under the CC-BY 4.0 licence.

Helmholtz resonator cells can be highly effective in tailoring new designs of acoustic metamaterials for a wide range of applications. However, a principal consideration in the design process is the practical fabrication and operation of the metamaterial concept.

4.3.4 MANUFACTURING TECHNIQUES AND CHALLENGES

Since acoustic metamaterials largely rely on their structure to generate the dynamic features of interest, rather than the material properties (Gardiner et al., 2021), much of the focus on innovations in acoustic metamaterials has been, and will continue to accelerate into the future, on advanced manufacturing methods. One approach of note is additive manufacturing, which can be used to realise a wide variety of complex architectures for a broad range of materials. In this section, a brief overview of a selection of particularly popular additive manufacturing approaches is provided, drawing some detail from the work of Gardiner et al. (2021), with the aim of guiding the reader towards some practical contemporary techniques for metamaterial and device design.

Given the outlined information provided thus far on controlling or tuning active properties in an active acoustic metamaterial, specifically the effective density and bulk modulus, the manufacturing approaches detailed in this section are therefore applicable. Specifically, an active acoustic metamaterial has the capacity to be influenced or controlled by an external trigger (Gardiner et al., 2021). The fabrication techniques which can be used in the fabrication of such metamaterials can be broadly classified as the following principal methods of additive manufacture.

1. Vat polymerisation through stereolithography
2. Extrusion and deposition
3. Powder bed fusion

The brief list shown is not exhaustive, and there are a few other manufacturing methods of note including photon polymerisation, but the main aim of this section is to provide a concise overview of each principal approach listed, to aid in future strategies for acoustic metamaterial fabrication. The first of these methods is a **vat polymerisation technique** called **stereolithography**. This is a relatively familiar technique to many engineers and is based around the fabrication of a structure using a photocurable resin or liquid. The critical component in a stereolithographic process is an ultraviolet laser beam which is used to cure each layer of liquid used as the manufacturing medium, where the metamaterial being assembled is then submerged to form the subsequent layer, after which stage the next layer can be deposited and cured with the aid of the laser beam. The laser is used to effectively *sketch out* the metamaterial structural configuration, as part of the general photo-curing step. It is possible to achieve micron-scale levels of precision in stereolithography (Gardiner et al., 2021), but the challenges associated with this technique are that the mechanical properties of the material can be

complex to tune and control, and that there can be a significant amount of processing required once a structure is printed.

An advantage of stereolithography is that it is suitable for the additive manufacturing of structures incorporating multiple materials, where structures consisting of multiple classes of resin have been demonstrated (Maruyama et al., 2020).
One possible benefit of this type of structure would be in optics, for example, in
the selectivity of signals travelling through part of a structure relative to another,
and, therefore, in associated sensor technologies. As an extension to stereolithography techniques, the additive manufacturing of a ferroelectric metamaterial has
also been demonstrated in the 2020s (Hu et al., 2020), using imidazolium perchlo-
rate as the base material. Using a stereolithographic approach, it has been demonstrated that this ferroelectric metamaterial can be fabricated to exhibit a tuneable
stiffness, and thereby a controllable sub-wavelength bandgap. There are many
exciting prospects for such technologies, providing the manufacturing challenges
already outlined in this book can be overcome. For industrial upscaling, improving the speed of fabrication should also be investigated, but this will undoubtedly
follow with the continued improvements to 3D-printing facilities.

Vat polymerisation is sometimes referred to as vat photopolymerisation, and as
a concept it more generally refers to the process by which resins are selectively
cured using light activation. Stereolithography is therefore specifically a sub-set
of the vat polymerisation principle. At their bases, all vat polymerisation methods
rely on the selection of a suitable resin to act as the base for the acoustic metamaterial. Light activation of a resin occurs when the liquid reacts to the wavelength
of light to which it is exposed, and the molecules within the resin then act to bind
together to synthesise the metamaterial structure. Another example of a vat polymerisation technique is digital light processing, where patterns of light are used to
define the structure of an acoustic metamaterial (Gardiner et al., 2021). In terms
of these techniques, many are still relatively time consuming and expensive for
large-scale industrial manufacture, and this remains a key barrier to wider exploitation. Some of the limitations on time, for example, can be attributed to the time
taken for resins to cure through vat polymerisation techniques.

The second general approach to additive manufacturing for acoustic metamaterials is that of **extrusion**, where many classes or configurations of material have
traditionally been synthesised, for example, metals, ceramics, plastics, and biological materials (Gardiner et al., 2021). At larger scales, extrusion-based techniques are commonplace, but much further development is required to achieve the
benefits for the fabrication of acoustic metamaterials on the small geometric
scales, but in large industrial volumes of manufacture. The sub-set manufacturing
approach of interest to this subject matter is fused deposition modelling. Whilst
there are a few other extrusion or deposition approaches available, including jet-
ting, fused deposition modelling is generally regarded to be a cost-effective and
relatively practical choice for acoustic metamaterial fabrication. It is sometimes
referred to as fused filament fabrication, and the basic principle is that a component is constructed in layers via the deposition and precise placement of melted
material along a programmed path. Due to the required nature of the material to

be deposited, it is common to use thermoplastic polymers. Challenges associated with the fused deposition modelling technique include ensuring the mechanical performance of the deposited material, configuring the cure time, and enhancing the precision of the deposition process.

Whilst there are hybrid approaches to acoustic metamaterial fabrication available and which can be further reviewed in the scientific literature (Gardiner et al., 2021), the final approach to fabrication that will be briefly outlined is the **powder bed fusion** technique. At a general level, powder bed fusion relies on a quantity of powder material to be processed, deposited over an area on which the acoustic metamaterial will be synthesised. A laser beam is then used to process the material and generate the first part of the acoustic metamaterial, before a new layer of powder is deposited and the process is repeated. There are several processes under the umbrella of powder bed fusion (Gardiner et al., 2021), and many exhibit challenges with regards to the fabrication of acoustic metamaterial structures incorporating different materials. A selection of subset techniques which can be classified as powder bed fusion approaches includes electron beam melting, selected laser sintering, and multi jet fusion. It is evident that there are several manufacturing opportunities for acoustic metamaterials, but much further development in the realisation of practical devices to complement the advances in mathematics and modelling is required. Some of this will be explored in the next section.

4.3.5 APPLICATIONS AND OUTLOOK

Traditional applications of acoustic metamaterials encompass the control or reduction of noise sources, the focusing or shaping of sound waves, for example, through custom fabricated lenses, in environments with specific limitations, such as space, and to overcome geometrical limitations. For example, acoustic metamaterials are particularly useful in biomedical ultrasound, where, for example, MHz range transducers can be fabricated using acoustic metamaterials to deliver a sufficiently small attenuation of the ultrasound signal in a relatively small space (Rojas et al., 2021). In such approaches, the acoustic metamaterial can be used to form the backing layer of a device, for example. The clear advantage of using an acoustic metamaterial is that there is a freedom in terms of the geometrical features which can be engineered, and if an additive manufacturing technique is utilised, then precise control of acoustic features can be achieved. In such biomedical ultrasound devices, it is possible to guide sound echoes from the device away from specific components in the transducer, such as the piezoelectric material, through careful definition of the acoustic metamaterial geometry (Rojas et al., 2021). If ultrasound can be manipulated in this way, and directed along specific defined paths, then the implication is that smaller and thinner devices are now possible, but able to perform with the dynamic characteristics of devices which would traditionally need to be much larger or thicker. It is expected that biomedical ultrasound devices such as these will continue to be innovated with acoustic metamaterials as central components to their design and manufacture. Technologies have already demonstrated their potential for the measurement of cardiac CT imaging (Rojas et al., 2021), and in 2021 an acoustic metamaterial was demonstrated for enhanced acoustic transmission for biomedical ultrasound towards the MHz range, by making use of negative refractive index (Wang et al., 2021). An example of an acoustic metamaterial designed and fabricated for a practical application, in this case, enhanced underwater ultrasound imaging, is shown in Figure 4.12, extracted with permission from the work of Kim et al. (2022).

In the results shown in Figure 4.12, the spherical and ellipsoidal lenses are two types of acoustic metamaterial engineered for the purposes of improving the frequency bandwidth and the achievable angle of measurement for underwater ultrasound imaging. The lenses were fabricated using stainless steel and additively manufactured, before being configured in an experimental setup as shown in Figure 4.12(a), with a hydrophone used for measurement (Kim et al., 2022). In general, these acoustic metamaterial lenses were demonstrated to show promise for improved resolution ultrasound imaging in underwater environments. For example, ultrasound imaging was achieved for the ellipsoidal lens across the range of 60 kHz

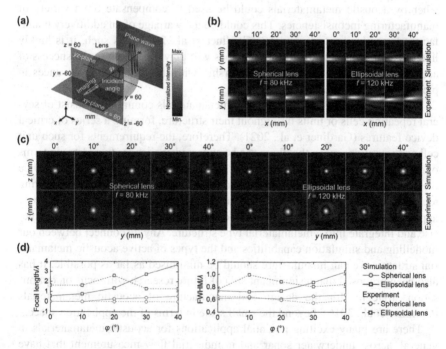

FIGURE 4.12 A three-dimensional acoustic metamaterial for enhanced underwater ultrasound imaging. Here, both finite element simulation (using COMSOL Multiphysics®, COMSOL, Inc.) and experimental results are shown, demonstrating (a) the experimental schematic, (b) and (c) ultrasound imaging for two types of lenses, and (d) focal lengths and full width at half maximum (FWHM) data for incident angle φ. Reprinted from Mechanical Systems and Signal Processing, Vol. 179, Kim et al., Three-dimensional acoustic metamaterial Luneburg lenses for broadband and wide-angle underwater ultrasound imaging, Page 109374, Copyright (2022), with permission from Elsevier.

to 160 kHz. The direction of innovation in future years will likely expand on these measurement capabilities. For example, there is significant potential for devices able to capture a range of measurements in the human body, in various configurations and conditions, and if geometry is no longer a real constraint, then this will open the possibility for a new range of novel and miniaturised devices for healthcare and wellbeing.

Acoustic metamaterials have recently been proposed for or utilised in the field of earthquake or seismic engineering, for example, to make use of bandgaps in a device to counteract the damaging effects of low frequency seismic vibrations, noise reduction systems (Gao et al., 2022), and underwater applications such as sonar (Gardiner et al., 2021). However, more broadly there will be a distinct and growing presence of acoustic metamaterials in adaptive ultrasonic devices. One of the main avenues of interest in terms of how transformative they may be for the device landscape is because they exhibit the potential to mitigate traditional limitations associated with geometry. It will be possible to create practical ultrasonic devices at length scales previously thought to be impossible, and configurations whereby acoustic metamaterials could be used to compensate for a variety of manufacturing inconsistencies. This could be the warpage of an additively manufactured membrane, for example, (Gardiner et al., 2021). As such, it is highly likely that additive manufacturing methods will play a key role in the success of active acoustic metamaterials, in partnership with the forthcoming advances in piezoelectric materials.

It has been established that acoustic metamaterials commonly consist of several repeating cells or units throughout their structure, forming a series of critical device features (Gardiner et al., 2021). Therefore, the requirements for such unit cell configurations mean that there can be expected challenges to address in terms of device fabrication consistency and repeatability. It is an aspect of device manufacture which will continue to be addressed through additive manufacturing techniques. Another major challenge in acoustic metamaterials for ultrasonic applications will be the compositions or classes of material we are able to synthesise and integrate into a metamaterial type structure. Any disconnect between our modelling and simulation capabilities and the types of active acoustic metamaterial structure we can manufacture should be minimised as far as possible. It has also been proposed that a significant area of future research will focus on optimisation strategies for the design and manufacture of acoustic metamaterials (Cummer et al., 2016), beyond the small scale in terms of the application space.

There are many exciting potential applications for acoustic metamaterials in general, across underwater sonar and in industrial flow measurement that have received little attention. Typical applications of acoustic metamaterials in adaptive ultrasonic devices in the future will likely include, but not be limited to, resonant agile sensors, precision proximity devices for flow measurement, energy harvesters capable of gathering energy in a broad range of modes, and even towards a new generation of advanced wearable device based on piezoelectric materials or ultrasound. There might also be the possibility of implementing devices of this nature with different functions, for example, to collect data relating to the health of a subject but also act as an energy harvester. Steerable and unidirectional wave emission has already been demonstrated by using shape memory alloys in meta-surfaces (Song and Shen, 2020), and such developments will likely continue. In general, the scientific literature, and technical experience, has shown several challenges which remain if we are to make significant future progress in devices with enhanced properties through acoustic metamaterials. A selection of the key challenges can be summarised as follows (Gao et al., 2022; Gardiner et al., 2021; Ma and Sheng, 2016; Zangeneh-Nejad and Fleury, 2019):

1. Expansion of the operational capabilities of acoustic metamaterials with regards to frequency and bandwidth.
2. Establishment of suitable manufacturing approaches to realise acoustic metamaterials of sufficient precision and *feature repeatability* in their structural configuration, especially with respect to the fabrication of relatively large materials with effectively *small-scale* embedded features, thus removing traditional geometrical constraints for tuning dynamic properties.
3. The integration of structural tuneability in the acoustic metamaterial, including in the manufactured condition.
4. The engineering of strategies for the computation or digitisation of meta-materials as binary units, which will likely be useful in the design and fabrication of metamaterials and metasurfaces for manipulating sound.

4.4 SUMMARY

If we can consider the material challenge for ultrasonic transducers relatively broadly, it is evident that the careful selection of materials in device design is critical, and this has been known for many years. The dynamic performance of ultrasonic transducers is highly dependent on key material properties including Young's modulus, the density, the acoustic impedance, and the geometrical size and shape. At the fundamental level, these principles are little different in the case of adaptive ultrasonic transducers, where again these material properties and configurations are critical to performance. The major challenge is that the fabrication approaches for shape memory materials and acoustic metamaterials, the principal classes of advanced material for adaptive ultrasonic devices, are highly complex, relatively expensive, and time consuming. It is for these reasons that continued innovations in advanced materials are vital, and where it is likely that additive manufacturing will play a central role in forthcoming years. As this technology matures, and we understand more about how to engineer or ensure optimal material properties using these methods, then the fabrication of both shape memory materials and acoustic metamaterials for adaptive ultrasonic transducers will become more practical. From this, a new frontier of multifunctional and intelligent ultrasonic devices will be within reach. Conventional approaches such as electrical discharge machining in the case of shape memory materials, whilst effective and functional, are too time consuming and arduous to make many classes of adaptive ultrasonic transducer feasible on an industrial scale. There is a similar associated challenge for additively manufactured acoustic metamaterials, where common fabrication approaches require high levels of precision and time to ensure the printing of devices with sufficient quality.

Nevertheless, the field of acoustic metamaterials, with particular focus on the active variety of this class of materials, presents an exciting future route for innovation in advanced materials and transducers exhibiting multifunctional and adaptive features. There are clear cases where an active acoustic metamaterial may be preferred over a shape memory material in the design of an adaptive ultrasonic device. One advantage is that there is the capacity to more readily or rapidly manufacture a device based on an acoustic metamaterial. There are still challenges around the time required and complexity of many shape memory materials, and so for the design of certain configurations of adaptive ultrasonic transducer, an active acoustic metamaterial approach may be the optimal route. However, additive manufacturing approaches can still be time consuming, and if a large volume of devices are required then the time and expense incurred for utilising acoustic metamaterials can accelerate. This is further complicated if a batch of devices should be designed to display similar dynamic features, where challenges around repeatability become significant. Another challenge with acoustic metamaterials is the achievable operating frequencies for the size of the device. Important advances have been made in this area, as described in Gardiner et al. (2021), but the limits of available fabrication approaches can often not overcome the physical challenges of transducer manufacture with the desired dynamic features. For example, many resins require specific post-curing processing, and it is not uncommon for fabricated parts to contain inhomogeneities or warped features, even on a small scale. In terms of the time taken in an additive manufacturing process, although many are still relatively slow, deposition or extrusion approaches are generally known to be faster than many of the other alternative methods available (Gardiner et al., 2021). In any case, fabrication approaches for shape memory materials, whilst not directly comparable, are also highly time consuming. For a large scale and rapid impact of advanced, adaptive ultrasonic devices, further progress to address the material challenge is required. Primarily, this will likely involve significant accelerations in precision manufacturing at extremely high rates and volumes, whilst maintaining an optimal quality of material fabrication, in terms of both feature definition and material properties.

5 The Road to Intelligence

5.1 SCOPE

The advances in adaptive ultrasonic transducers outlined thus far in this book have encompassed a historical perspective of transducer design, fabrication, and optimisation, a discussion of key principles associated with the design and construction of a few popular classes of transducer, and a comprehensive account of advanced materials for transducer design, as detailed in Chapter 4. It has been shown that the design of an adaptive ultrasonic device can be a complex and time-consuming process, with significant mathematical modelling and simulation demands. It is evident that it will take time for the adaptive ultrasonics discipline to progress in a significant and impactful way into the application space and beyond the laboratory, and whilst part of this is due to the manufacturing challenges, the complexities around integration with modern applications is another consideration. The potential capacity of adaptive ultrasonic transducers to enable a wide range of multifunctional and tuneable device features is significant for practical applications. As was outlined in Section 1.5, there are critical emerging technology demands associated with device tuneability and resilience, as we embark on transducer development for more complex and challenging environments. The major aspect we have not yet considered in much detail is the progress towards the integration and application of intelligent features in an ultrasonic device, whether they be inherent to the device performance, for example, dynamically or through another physical mechanism, or related to the application.

The focus of this chapter bridges a few key themes. First, a general overview of some approaches currently used to integrate or exploit the capabilities of artificial intelligence in ultrasonic devices and their applications is provided, with a general review of current progress in both industrial and medical ultrasound fields. The second part of this chapter provides some key information on adaptive signal processing strategies which can be implemented, from beamforming with array-type systems to adaptive signal shaping. This includes details of the time-reversal method as a key example of how technologies can be engineered for complex environments and those where the properties can change. Finally, the chapter concludes with the proposition of some likely strategies that may become more common as the field of adaptive ultrasonics continues to advance. Whilst the design, development, and implementation of advanced materials in ultrasonic devices, as detailed in Chapter 4, is central to future innovations in adaptive ultrasonic devices, it is also critical to progress our engineering capabilities in signal processing. Through design strategies combining aspects of materials science with

DOI: 10.1201/9781003324126-5

computational oriented parameters, it will be possible to realise technology platforms with the adaptive features necessary for a new generation of multifunctional and tuneable ultrasonic device. Technology advances from the perspective
of materials science should therefore align with the necessary developments in
signal processing and artificial intelligence where possible.

5.2   ARTIFICIAL INTELLIGENCE AND ULTRASONICS

Artificial intelligence is rapidly receiving greater attention in the scientific literature, and the public consciousness, but it is arguably an extremely broad concept that can take different forms depending on the applications of interest. It
is not unreasonable to predict its pervasiveness in many new disciplines of science in technology in forthcoming years, but before we consider it in the contexts of ultrasonic devices and their applications, we can clarify its meaning. The
Oxford English Dictionary defines artificial intelligence as the following (OED
Online, 2024):

> *'the capacity of computers or other machines to exhibit or simulate intelligent be-
> haviour; the field of study concerned with this. In later use also: software used to
> perform tasks or produce output previously thought to require human intelligence,
> esp. by using machine learning to extrapolate from large collections of data'.*

Therefore, in the context of ultrasonic technologies, any feature of artificial intelligence embedded in a system should have some element of decision making or
analysis associated with it. Artificial intelligence is a technology that has rapidly
advanced in the 21st century, and it is constantly finding new applications in a
multitude of fields, across science, medicine, and industry. The principle of artificial intelligence is sometimes used interchangeably with machine learning, but
they are technically distinct from one another. Using Oxford Reference (Oxford
Reference, 2024), machine learning can be regarded as:

> *'a branch of artificial intelligence concerned with the construction of programs that
> learn from experience. Learning may take many forms, ranging from learning from
> examples and learning by analogy to autonomous learning of concepts and learning
> by discovery'.*

Therefore, one distinction which can be made is that artificial intelligence
approaches should in some way simulate intelligent behaviour consistent with that
expected of a human (Malik et al., 2019), whereas a machine learning approach
does not need to because it only should learn from a given experience, based on
the Oxford Reference descriptor provided. The key question for adaptive ultrasonics, and ultrasound as a discipline more generally, is how is the rapid growth and
uptake in artificial intelligence influencing the field of ultrasonics? One of the
aims of this chapter is to provide some detail of the status in ultrasonics regarding
artificial intelligence, and likely future pathways to implementation and application. For the purposes of this part of the book, machine learning will also be

referred to in the discussions relating to artificial intelligence, being a subset of this concept as defined above.

It is important that artificial intelligence is considered in the context of adaptive ultrasonic devices. The principal reason for this is that artificial intelligence can reasonably be seen as a key enabler in how the performance parameters of ultrasonic devices and their systems can be controlled. For example, if an ultrasonic transducer can be designed to operate in multiple vibration modes with a shape memory material, consistent with the approaches set out in Chapter 4, then it would be feasible to propose a system that could learn how to respond to a change in an environmental condition, and to adjust device performance accordingly. This may be the resonance frequency, or even the physical shape of an end-effector or the configuration of the transducer itself. Another way in which artificial intelligence approaches can be used in ultrasonics is through the processing of data, for example, in the discipline of ultrasound imaging. More information on this is provided later in this chapter, but artificial intelligence has the potential to optimise the processing and analysis of imaging data which could, for example, revolutionise patient outcomes. Key distinctions in these examples are how artificial intelligence approaches are administered. For instance, in the latter case, artificial intelligence can be used as part of a data post-processing strategy, and there is not necessarily a requirement to be consistent with the principles of adaptive ultrasonic technology. However, for the former example, it is conceivable that artificial intelligence methods could be reliably integrated with the design and operation of ultrasonic devices in the future, thereby forming a class of adaptive ultrasonic technology platform. In an example from 2023, Gandomzadeh et al. report the application of machine learning approaches including neural networks to process experimental data in the optimisation process for the design of a magnetostrictive ultrasonic transducer (Gandomzadeh et al., 2023). It is an example of a study where real design parameters for an ultrasonic transducer are selected for optimisation using a machine learning approach, with a subsequent optimised configuration proposed. In this case, design parameters for the transducer included the current through the magnetostrictive coil and aspects of the transducer horn geometry (Gandomzadeh et al., 2023). The design parameters were varied and processed using the machine learning approaches, including the radial basis function form of neural network, where optimisation of performance could be assessed based on parameters such as output amplitude (Gandomzadeh et al., 2023). As an illustrative case of what an optimisation protocol may look like and to encourage the reader to investigate further in this regard, an example schematic of a neural network employed by Gandomzadeh et al. is depicted in Figure 5.1, showing the range of design parameters investigated on the left side of the image, in the designated 'input layer' (Gandomzadeh et al., 2023).

Here, the input layer of the artificial neural network, as shown in Figure 5.1, can constitute the physical input conditions for an engineering system, in this case a magnetostrictive ultrasonic transducer. There are then a series of 'hidden layers', which can be considered as the training layers for determining an optimal output, in this case the amplitude of the transducer, noted as $Am.$ by Gandomzadeh

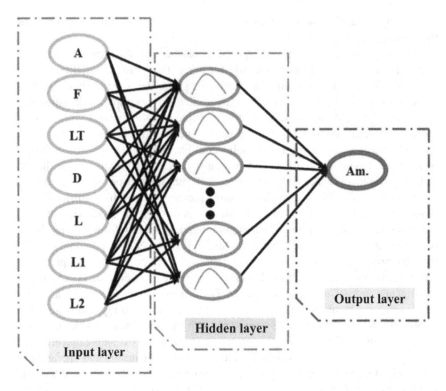

FIGURE 5.1 A schematic of a neural network used in the optimisation of the design for a magnetostrictive ultrasonic transducer, through a machine learning approach, extracted from the work of Gandomzadeh et al. (2023). The lettered items in the neurons in the input layer can be attributed to separate design parameters for the magnetostrictive ultrasonic transducer. Reprinted (adapted) with permission from Gandomzadeh et al. Copyright 2023 American Chemical Society.

et al. and which is also shown in Figure 5.1. Artificial neural networks like this are effectively training processes for the input data and they can be computationally demanding, involving many cycles of training through hidden layers. Furthermore, weighting factors are typically applied to the data, including between the hidden layers and the output layer (Gandomzadeh et al., 2023), especially as more is learned about the data sets. For example, this could be for a property or factor which makes more of an influence on the output property, in this case, the amplitude. Gandomzadeh et al. note that the optimal neuron count for the hidden layers is complicated and non-standardised (Gandomzadeh et al., 2023), where it is reported that many studies still rely on time-consuming and computationally intensive trial-and-error approaches. Mathematical methods including root-mean or sum-of-squares error metrics have been reported as viable pathways to determine an optimal number of hidden layers in an artificial neural network like this (Gandomzadeh et al., 2023).

If we look a little more broadly at how artificial intelligence approaches have been used in ultrasonic technology over the previous few years, and more generally across industrial and medical ultrasonics, we can reveal a more detailed picture of technology trends and where future activities may be progressing. Like for any complex field with a detailed history over many years, it is only feasible to provide a selection of highlights in progress, with key illustrative examples of technology developments as relevant to adaptive ultrasonics. Here, we can clarify some of the minor differences between artificial intelligence, machine learning, and this time considering deep learning. Shen et al. provided a comprehensive overview of how artificial intelligence is used in ultrasound (Shen et al., 2021), where the key concepts were differentiated, including artificial intelligence (where human cognition is imitated and includes machine and deep learning techniques), machine learning (the mapping of inputs into outputs, with elements of training), and deep learning (multiple artificial neural networks). Shen et al. also noted the convolutional neural network as a stand-alone subset and consisting of images, which is important for disciplines such as biomedical ultrasound. There has been a wealth of studies focused on applying artificial intelligence techniques in the processing and reconstruction of images in biomedical ultrasound, with a selection of notable examples across obstetrics, thyroid, and the musculoskeletal system, among others, as detailed in the work of Shen et al. (2021).

Despite the popularity of artificial intelligence for ultrasound imaging and processing, there is evidence of significant enhancements to many non-medical imaging ultrasonics applications using such methods, and these will be briefly reviewed in this chapter. For example, in a 2011 issue of the *Journal of Nondestructive Evaluation*, a study reported a protocol for the classification of defects using the ultrasonic pulse-echo technique, through the implementation of an artificial neural network (Sambath et al., 2011). This is a relatively early but important example of the implementation of an artificial intelligence approach for an industrial application. The primary objective of this study was to properly classify defects obtained using pulse-echo ultrasound, across a series of designated flaw types (in this case, this was porosity, a lack of fusion identified, any tungsten present, and a non-defect condition). The rationale behind the study was that defects tend to be processed and understood by a skilled human operator, and so by translating this knowledge to an artificial neural network, the procedure could be automated, and potentially rapidly accelerated. Progress in the application of artificial neural networks had already been made around this time (Sambath et al., 2011), but many ultrasonic applications were still exploratory or theoretical. The basis of an artificial neural network is to establish a training regime, effectively a machine learning method, to teach a system how to process data and identify desired features or parameters (Wu and Feng, 2018). The major contributions of the research of Sambath et al. were to demonstrate how an artificial intelligence approach could feasibly be used to detect a range of defects for an ultrasonic nondestructive testing and evaluation process. Readers are encouraged to consult expansive articles and reviews such as that by Uhlig et al. (2023), which provide highly detailed accounts and analyses of how artificial intelligence techniques have been applied in the field of nondestructive testing and evaluation. Fundamentally, as Uhlig et al. indicate, and consistent with the perspectives of others in the field, a key challenge in the implementation of artificial intelligence techniques for enhancing the speed, efficacy, or performance of nondestructive testing and evaluation processes is the availability of data in terms of both volume and quality.

A major step forward was demonstrated in 2024 for ultrasound imaging of the lung, where only around 300 images were shown to be required to facilitate the deep learning training process for the monitoring and measurement of lung health (Howell et al., 2024). This shows a promising trajectory for how training protocols can be developed in future, though it is arguable that some imaging or processing applications may require fewer images or quantities of data for training compared to others. Uhlig et al. noted the difficulty of obtaining high-quality experimental data for use in the training protocols, and therefore the development of synthetic data has become more popular, using a process referred to as data augmentation. This is where computational methods can be used to create artificial datasets large enough to undertake the necessary training. It has also been noted that there is a real challenge with ensuring the quality of output in a physical sense, because we must ensure that we are working with models and systems that are realistic (Uhlig et al., 2023). Research into the application of artificial intelligence techniques for nondestructive testing and evaluation has extended into that of detecting rail defects (Li et al., 2024), and for the reconstruction of images enhanced by deep learning, for ultrasonic imaging using transducer arrays which generate guided waves (Wang et al., 2024), where in the latter it was notable that resolution was identified as a continuing challenge.

A key study of note to ultrasonic engineers with interests in ultrasonic welding, an application commonly incurring high power levels or subjecting an ultrasonic device to relatively high levels of loading, was demonstrated in the *International Journal of Advanced Manufacturing Technology* in 2012 (Norouzi et al., 2012). There are long-standing challenges with optimising process parameters for ultrasonic welding which are reported in detail in the literature, where the work of Khan et al. (2017) and Singh et al. (2017) are notable examples. Process parameters in ultrasonic welding can include factors such as time to achieve a weld, the ultrasonic amplitude applied to the substrate, and the applied pressure from the welding system. In this case, Nourouzi et al. made use of artificial intelligence approaches including the implementation of artificial neural networks to facilitate the selection of an optimal batch of process parameters for a welding application, with a reported calculated strength in the order of 10% higher than an equivalent but non-optimised condition. A notable study by Li et al. followed in 2020 which used artificial intelligence techniques including artificial neural networks to predict the quality of welds for carbon fibre–reinforced thermoplastics (Li et al., 2020). The reader is also invited to consult a review which is sufficiently broad in scope regarding the use of artificial intelligence for the assessment of weld defects using nondestructive testing and evaluation techniques (Sun et al., 2023).

It is interesting that developments in the application of artificial intelligence approaches in the field of ultrasonics progressed for both traditional power ultrasonic applications and those for nondestructive testing and evaluation, across both industry and medicine. In one interesting study, Korolev et al. demonstrated the use of artificial intelligence techniques for assessing the cavitation behaviour of water-based liquids containing alcohol, exposed to ultrasound (Korolev et al., 2022). Here, artificial intelligence was used to assess a series of images gathered by experiment, with statistical patterns drawn from them. In this case, it was found that artificial neural networks were of particular use in the differentiation of images according to the associated concentration of alcohol in that water-based solution. In 2019, van Sloun et al. delivered an in-depth analysis of how deep learning has been integrated with ultrasound imaging technologies (van Sloun et al., 2019). Here, van Sloun et al. discuss the need for imaging of the highest quality and expediting rapid representative images from a procedure are evidently drivers for implementing advanced deep learning and artificial intelligence approaches in ultrasound. A key example of an ultrasound imaging process enhanced using an artificial intelligence approach is where the process of ultrasound localisation microscopy was combined with a deep learning technique, as reported by van Sloun et al., in the use of a neural network to transition an input image of low resolution, to that of a high-resolution output. The fundamental innovation is the delivery of sufficiently high-resolution images for the high-frequency imaging being undertaken, often in the MHz frequency range. It should be noted that van Sloun et al. are reviewing the primary contributions of Khaing et al. (2018) in this case, as the key innovators of this technique (Khaing et al. 2018). Fundamentally, important considerations for how deep learning approaches such as neural networks should be implemented in ultrasound, as summarised by van Sloun et al., include factors such as the volume of data required to properly train a system, and the availability of the useful data needed to ensure that neural networks can produce the necessary conditions for optimisation, and it is noted by van Sloun et al. that such challenges are of relevance outside that of ultrasound imaging. This is of little surprise, given the challenges and complexities associated with various ultrasonic applications, whether they be low or high power, or destructive or nondestructive in their nature. As an illustrative example of how artificial intelligence techniques are being applied in ultrasound imaging, an example from the clinical field in the imaging of an elbow for joint pain is shown in Figure 5.2, extracted from the work of Inui et al. (2023).

Here, Inui et al. show how deep learning can be used to train a user interface to recognise instances of osteochondritis dissecans, a type of pain which can affect joints such as the elbow. The training regime applied to the system showed significant promise, where there was a relatively high reliability in the image classification and detection of osteochondritis dissecans (Inui et al., 2023). Techniques such as this illustrate the real potential of embedding artificial intelligence in ultrasonics to enable rapid clinical processing.

The objective of this section has been to set out a series of notable developments in the application of artificial intelligence to ultrasonics. There has been

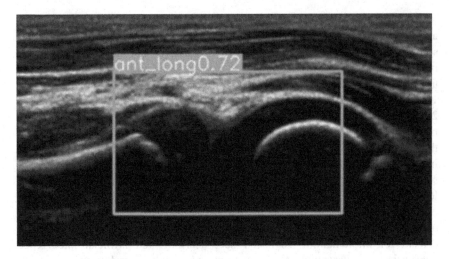

FIGURE 5.2 An example of ultrasound imaging in the biomedical field, showing the detection of osteochondritis dissecans in the elbow, based on an artificial intelligence approach. The overlay frame is part of the graphical user interface used in the approach. Reprinted (adapted) from the work of Inui et al. (2023), under the CC-BY 4.0 licence.

significant activity in ultrasound imaging, and part of the reason for that is likely because of the modes of training available. If an image is an output of an ultrasonic process, for example, by experiment, then it is practical to use that image to train a computational mechanism, such as an artificial neural network, to deliver a system that is more readily able to detect such features in the future. The other aspect of intelligent technology in relation to ultrasonic devices is on the signal processing front. With the advent of array technologies in recent years, and complex avenues of measurement and sensing, it is important we can achieve flexibility in how ultrasound can be focused and directed. With a new generation of advanced technologies, it is important that we can tailor system response, for a series of high-performance acoustic and ultrasonic devices. The next section focuses on a few adaptive signal processing strategies which have become popular, with an overview of common applications and how they can be practically implemented.

5.3 ADAPTIVE SIGNAL PROCESSING AND BEAMFORMING

Many contemporary applications of ultrasound, for example, in biomedical imaging, require the use of arrays of transducers to generate a series of ultrasound beams which can be focused or directed in a particular path or direction, or targeted at a specific area in a subject material. The use of array formations is necessary because time delays can be configured into the system, to establish a directionality to the propagation of ultrasound. There are many useful and detailed reports which provide a comprehensive foundation to the generation of ultrasound wave fields and how these can be manipulated, and so the intention here is to deliver a broad overview of wave field generation and how this relates to the emergence of adaptive ultrasonics.

To conclude with an overview of some adaptive beamforming strategies which are finding success in several applications, including biomedical imaging and nondestructive testing and evaluation, the principles of beamforming in materials should first be established. David et al. provided a concise overview of beamforming approaches for homogeneous media in their research investigating compressive beamforming in the time domain (David et al., 2015). Here, an array of stand-alone ultrasonic transducers can be configured in an array formation, where each is designated as a separate channel that can be triggered at a set time. A representation of this is depicted in the schematic shown in Figure 5.3, extracted from the work of Mozumi and Hasegawa (2018), and which demonstrates the influence of different approaches to beamforming on the quality of ultrasound images (Mozumi and Hasegawa, 2018).

The main outcome of the study used to produce the data shown in Figure 5.3 was to show how adaptive beamforming can be used to enhance the quality of ultrasound images (Mozumi and Hasegawa, 2018). In general, the element size, or the transducers and their relative spacing, are critical to the performance of an array (David et al., 2015), where a similar ultrasound pulse can be generated across all the transducer elements but triggered at different times, in a delay

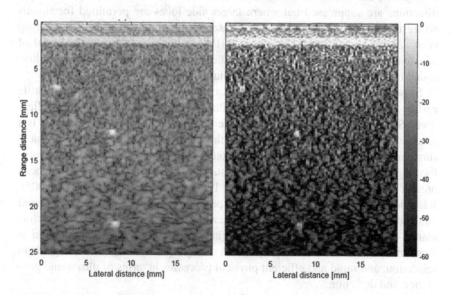

FIGURE 5.3 Images showing the influences of different methods of beamforming on the quality of ultrasound imaging on a phantom, where the left image has been produced using a classical delay-and-sum approach, whereas adaptive weighting has been used to produce the image on the right. Reprinted (adapted) from the work of Mozumi and Hasegawa (2018), under the CC-BY 4.0 licence.

approach. This means that it is important to know the time of flight (often commonly referred to as ToF in the literature), and as David et al. have indicated, high-quality ultrasound images can be generated by the combination of returned data from a series of focused ultrasound pulses which are emitted in several directions (David et al., 2015). It is also noted that, customary for nondestructive testing and evaluation and ultrasound imaging processes, the reduction or elimination of noise is important, and such focusing approaches can be highly effective in this regard.

In general, a series of discrete points in a medium can be designated with two components, such as x and y, from where the ultrasound intensity from the reflections can be calculated using the ToF for each transducer or channel (David et al., 2015). Using this information, as David et al. note, delays can hence be calculated to establish a case of constructive interference from a particular x, y point in the target medium, in a process David et al. refer to as Delay-and-Sum (DAS). This DAS process is also noted in the work of Mozumi and Hasegawa and is referred to in Figure 5.3. This approach has been noted to be a commonly applied technique in ultrasound imaging for medical applications (Synnevag et al., 2007). The principle of DAS can be further extended to demonstrate the principle of *adaptive beamforming*, with a robust example demonstrated in the work of Synnevag et al., which investigates the approach for imaging in medical ultrasound applications. Here, it is shown by Synnevag et al. how adaptive beamforming instead employs a weighting calculation to the system, where ultrasound signals in non-principal directions are suppressed but where larger side lobes are permitted for signals where there is no reception by the ultrasound measurement system (Synnevag et al., 2007; Mozumi and Hasegawa, 2018). In this way, the principal goal of adaptive beamforming can be realised, which is to enhance the resolution of ultrasound images gathered from a target medium.

Sasso and Cohen-Bacrie demonstrated a complete adaptive beamformer a little earlier, in 2005, through their work on ultrasound imaging for medical applications (Sasso and Cohen-Bacrie, 2005). Here, it was acknowledged that DAS is a relatively straightforward approach, but that there can be implications on measurement relating to noise. It was also noted that adaptive techniques for beamforming are not entirely new, even at this stage in 2005, where concepts have been present in the scientific literature since the 1960s (Sasso and Cohen-Bacrie, 2005). It is evident that much of the literature, especially in the 21st century, has focused on the development and optimisation of new techniques for improving the performance of adaptive beamforming techniques, with key contributions tailored at signal noise reduction and the acquisition of high-quality images with regards to resolution, and rapid and efficient physical processes in terms of ultrasound generation and detection.

This section of the book is intended to provide a general overview of some commonly applied signal processing techniques, including adaptive beamforming, with examples and highlights of applications, rather than a chronological account of development. However, a comprehensive review by Zhang and Harvey in 2012 detailed some of the key approaches in signal processing for ultrasound, with a focus on nondestructive testing and evaluation for multilayered structures (Zhang and Harvey, 2012). Here, the authors showcased relevant techniques including averaging, cross-correlation, deconvolution, and filtering. The authors also demonstrated the primary motivations for introducing such signal processing techniques, including the accurate and reliable characterisation of defects in a material or structure, the elimination of noise or the improvement to signal-to-noise ratio, the acquisition of defect size and position within a structure, and the classification of signals (Zhang and Harvey, 2012). A key example from the scientific literature relating to an adaptive approach to processing ultrasound signal data in nondestructive testing and evaluation is from the work of Li and Hayward, which focused on adaptive array processing for enhancing the quality of imaging data (Li and Hayward, 2019). As a broad overview of the approach employed by Li and Hayward, a target location is specified on the medium under measurement, which requires appropriate delays to be implemented. Then, a pre-focusing step is instigated, where the system data prior to it being focused is processed before it passes through the beamformer algorithm. The final step constitutes the calculation of a weighting factor based on the contributions to the response from the signals that are in the direct axis of measurement, compared to those that are off axis (Li and Hayward, 2019). A series of beamforming approaches are detailed by Li and Hayward, beginning with the standard DAS technique as discussed. Others of note that are included by the authors include *minimum variance distortionless response*, which minimises the output power of the transducer array but ensures a sufficiently high input, and a *maximum likelihood estimator*, where the noise in the system has an assumption that it is stationary.

The need for adaptive approaches in signal processing extends beyond applications involving transducer arrays and those studies limited to the implementation of beamforming. The broad application of signal processing to a wide variety of scenarios includes that of individual or independent transducers, and it has been noted that in sensor technologies, such as biosensors, there is a significant need for both flexibility and adaptiveness for signal processing approaches (Pelenis et al., 2019). For example, the research of Pelenis et al. has demonstrated a biosensor based on a capacitive micromachined ultrasonic transducer; a convolutional neural network was employed to optimise the delay parameters, primarily to reduce the noise in the measurements (Pelenis et al., 2019). This is another example of an advanced signal processing strategy to mitigate the influence of noise on a system, and the authors compared the approach making use of a neural network algorithm to that of an adaptive filter which was implemented in prior research. The work highlighted the relative success of the neural network algorithm to improve the signal-to-noise ratio, showing the importance of such techniques for enhancing the detection performance of another common form of ultrasonic transducer.

Many of the ultrasound imaging applications discussed up to this point, with the techniques that can be implemented to generate images of sufficient resolution, can effectively be considered as inverse problems, where measurements or data can be used to determine the fundamental phenomenon. Reconstruction of

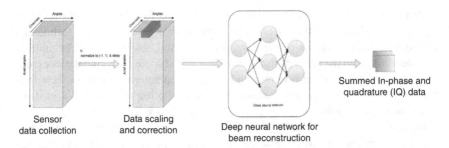

Sensor   Data scaling
data collection   and correction   Deep neural network for
                                                    beam reconstruction

Summed In-phase and
quadrature (IQ) data

FIGURE 5.4 An example of the process developed by Dahan and Cohen, where sensor information is obtained (the first block on the left) before being scaled (noted by the second block), followed by the input of the data to a neural network (designated as the third block from the left) for reconstruction of the summed data (shown by the block on the right). Reprinted (adapted) from the work of Dahan and Cohen (2023), under the CC-BY 4.0 licence, where the annotations have been enhanced.

images in ultrasound is a common inversion approach. As an extension from the concept raised in the work of Pelenis et al. regarding the implementation of a convolutional neural network, it is evident that deep learning approaches in adaptive beamforming have gained traction. For example, Luijten et al. presented a comprehensive account of how deep learning techniques can be applied for adaptive beamforming in ultrasound imaging (Luijten et al., 2020), with the notable achievement of undertaking adaptive beamforming using an artificial intelligence inspired technique that requires comparatively little training data to implement. This is significant, because for wider implementation in practical applications, it is imperative that techniques can be both rapid and inexpensive. As an illustrative overview, a representative platform innovated by Dahan and Cohen is shown in Figure 5.4 (Dahan and Cohen, 2023), in their development of an adaptive beamformer optimised using deep learning.

As an overview of the platform proposed by Dahan and Cohen, a convolutional neural network is used to enable the reconstruction of an image at multiple angles, for only one angle of measurement acquisition (Dahan and Cohen, 2023). Enhanced and enabled by artificial intelligence, this has interesting and significant potential for rapid measurement and inspection in the field of ultrasound imaging, especially if the training can be delivered in such a way as to ensure confidence in the data. Since convolutional neural networks are particularly effective at processing images, returned data can be suitably processed by the deep learning algorithm which can be adaptively tuned based on the set conditions on the algorithm. Fundamentally, the combination of an adaptive beamforming approach with a convolutional neural network has demonstrated the capacity to perform well compared to contemporary approaches, especially in terms of the image resolution that is generated, but with a significantly accelerated processing time and image reconstruction speed (Luijten et al., 2020). This is yet another example of the innovative use of artificial intelligence approaches in ultrasonics.

In a later example of the application of adaptive beamforming in ultrasound imaging, a technique for phase aberration correction was proposed by Mozaffarzadeh et al., to improve the accuracy of reconstructed images from collected ultrasound signals (Mozaffarzadeh et al., 2020). Here, imaging was undertaken to determine a comprehensive map of the sound speed associated with the target medium, with a representation of the necessary phase compensation determined using an adaptive form of the DAS beamformer. Regarding the implementation of the beamformer, certain physical quantities of the system were required, including the thickness of the target medium so that the phase compensation could be undertaken by the beamformer. As a mathematical representation of the beamformer developed and used by Mozaffarzadeh et al., the relevant DAS relationship employed in the system development is shown by Equation (5.1), which is extracted from the work of the respective authors (Mozaffarzadeh et al., 2020).

$$y_{\text{DAS}}\left(k\right) = \sum_{i=1}^{M} x_i\left(k - \Delta_{i,k}\right) \tag{5.1}$$

As a general overview of the parameters shown in Equation (5.1), $y_{\text{DAS}}(k)$ represents the data that is beamformed, k denotes the time index associated with the data, M represents the number of elements that are present in the array, and the $x_i(k)$ and $\Delta_{i,k}$ parameters are indicative of the received signals and the associated time delay, respectively, where i is the detector (Mozaffarzadeh et al., 2020). In a further study considering the application of adaptive beamforming for enhancing the quality of ultrasound imaging, Al Mukaddim et al. proposed an adapted configuration based on sub-aperture processing (Al Mukaddim et al., 2021), aimed at improving the image data in cardiac photoacoustic imaging, which is known to be particularly challenging in terms of ensuring the reconstructions of the images are of sufficiently high resolution. The basic overview of the approach employed, and as reported, by Al Mukaddim et al., involved the processing of two reconstructed images obtained via the DAS technique. Here, the channel information was divided into non-overlapping areas, after where the weightings could be calculated and reconstructed with optimally low levels of side lobes in the data (Al Mukaddim et al., 2021). A DAS beamformer is employed alongside beamforming processes for two sub-aperture data subsets which directly lead from the channel data. Then, a correlation and weighting determination is made, where the DAS beamformer algorithm is used to provide an estimate for the imaging from the photoacoustic sub-aperture processing. In effect, the weightings are based on a calculation determining how similar two images are from the sub-aperture data which is collected, and thus directly contributes to the reduction in the levels of the sidelobes (Al Mukaddim et al., 2021). As a general illustration of the effects of different beamforming strategies, particularly with reference to sub-aperture focusing, a range of pressure amplitudes obtained via beams which are unfocused, focused, and steered are shown in Figure 5.5 (Demi, 2018).

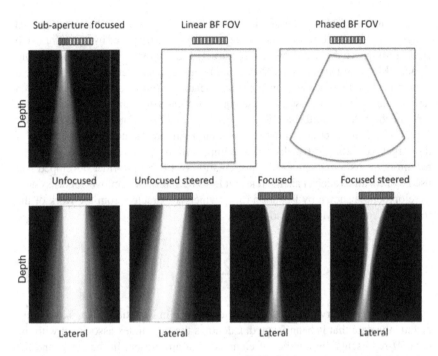

FIGURE 5.5   A range of beamforming (noted as BF in the images) outputs, showing the breadth of unfocused, focused, and steered conditions achievable. Note the field of view (FOV) acronym, and the inclusion of both linear and phased array variants of ultrasound generation. Reprinted from the work of Demi (2018), under the CC-BY 4.0 licence.

Another variant of the adaptive beamforming technique was demonstrated by Paul et al. around the same time, in their work on noise adaptive beamforming for linear array photoacoustic imaging (Paul et al., 2021). It is notable that whilst the DAS approach is frequently referenced, there remain concerns regarding the comparatively low resolution which is achievable compared to some of the more advanced techniques which have become more popular or preferred in the reconstruction of ultrasound imaging data. In the research of Paul et al. (2021), a novel adaptive form of weighting factor was proposed, called the *variational coherence factor*. The basis for this approach is that the coherence factor applied to the processing would account for variations in noise level. The frequency ranges studied in this research were in the order of MHz, and some images of suitably high resolution were generated using the approach. The variational coherence factor itself was established on using the mean and standard deviation of a series of delayed signals, depending on position (Paul et al., 2021).

Adaptive optimisation as a concept for processing imaging data from ultrasonic phased array systems has been reported to be successful for inspecting relatively large structures, and it has thus gained momentum for the analysis of a wide range of industrial systems. One example is the inspection of composite panels used in the construction of wind turbine blades (Duernberger et al., 2022). In this example, Duernberger et al. report the application of a multi-aperture adaptive beamforming approach in the inspection of composite panels for wind turbine blades. The size and spacing of the necessary apertures for analysis, where each aperture is an element or a group of them, are both calculated using the decay profiles of the ultrasonic signals (Duernberger et al., 2022). The authors report that for each combination of transmit and receive data, effective images can be reconstructed using a full matrix capture approach, taking account of ultrasonic amplitude. This approach is noted for its suitability for incorporating robotic modes of inspection, thus combining adaptive techniques with a level of automation such that relatively large structures can be inspected, whilst also maintaining speed of processing and a desirable level of spatial resolution (Duernberger et al., 2022).

The expanse of recent research has continued to address imaging challenges across industrial and medical applications, and it is evident that the principal factors for improvement associated with image reconstruction include the speed at which imaging data can be processed, the resolution of the images produced, and the contrast of the images such that the target features for analysis are prominent. Adaptive beamforming has remained a popular approach to the enhancement of ultrasound imaging, and there is evidence that the implementation of deep learning approaches to optimise the imaging process will continue (Wang et al., 2022). A key example is using neural networks to convert images of low resolution to those that are higher, such that the adaptive beamforming process can be optimised, such as using a minimum variance technique in a suitable form similar to those outlined (Wang et al., 2022).

5.4 STRATEGIES FOR INTEGRATING INTELLIGENCE

As a general summary, adaptive beamforming and signal processing strategies are continuing to gain traction for a range of measurement applications, both medical and industrial. However, it is apparent that there are opportunities for artificial intelligence techniques to be applied in the field of ultrasonics, beyond conventional imaging. One example is introducing some form of true decision making as ultrasound scans are being generated. Currently, artificial intelligence techniques facilitate high-quality images or faster analysis, but it may be possible to automate the selection of images in such a way that illnesses or afflictions (in the case of medical ultrasound) or specific defects (in the case of industrial ultrasound and nondestructive testing and evaluation) are more rapidly and effectively identified for beneficial outcomes. One of the next steps may be the adaption or selection of the artificial intelligence algorithm applied to the beamforming process, to optimise the generated and detected signals, and bypass the need for selection or application from a human user until the measurement process is complete. This may incur a long development time, to understand the limits of systematic processing and analysis. The introduction of true decision making in ultrasonic devices, not just at the processing level but also in the operation of devices themselves, would represent a significant step forward in realising highly intelligent ultrasonic technologies. Without such decision making, adaptive ultrasonics will remain more akin to a *smart* technology, and not one exhibiting the *intelligence* we may need in future.

Another strand of research inspired and enhanced through artificial intelligence is concerned with engineering design. There is already some development in this space, using artificial intelligence techniques to enhance or optimise design features of a variety of engineering structures based on a set of defined parameters or design goals. A key commercial-level example is the *Kosmos Torso One* from EchoNous (Redmond, WA, USA), which is effectively an ultrasonic diagnostic probe which uses artificial intelligence in its operational system to enable optimal scan times, image resolution, and probe placement (EchoNous, 2024). Improvements to patient recovery have been proposed, noting the potential benefits of ultrasound technologies incorporating artificial intelligence features. Regarding engineering design more broadly, a comprehensive review of the implementation of artificial intelligence techniques in the engineering design process was published by Yüksel et al., demonstrating how such algorithms can be applied to a range of engineering design problems, including, as examples, the design of bat-wing topologies and the shape or profile of aerofoils (Yüksel et al., 2023). As the authors identify, parts of these strategies include data-informed mechanisms to provide feasible solutions based on engineering design outcomes already known. Therefore, it is imperative that engineers and stakeholders continue to develop and store high-quality and robust data to inform the next generation of technologies. The concepts explored in the scientific literature regarding engineering design informed by artificial intelligence, can hence be applied in principle to the design of adaptive ultrasonic transducers. As shown throughout this book, there are a wide range of design considerations to incorporate into the development process for a transducer, irrespective of the chosen configuration. Depending on the modal and operational parameters of interest for the design, there are a myriad of physical parameters and properties to account for in the design process, including mass density, geometry, electrical characteristics, pressure, and temperature. The more data and knowledge we have for such physical parameters, then the greater and more numerous the opportunities will be for innovating future adaptive ultrasonic device concepts.

Of course, as we employ artificial intelligence as a design tool to greater extents across industry, there are risks. Some of the disadvantages of employing artificial intelligence techniques in engineering design have been summarised by Yüksel et al. and are included here for reference (Yüksel et al., 2023). First, it has been identified that a design solution developed using artificial intelligence may only be relevant for specific cases, and which may not be suitable for broader application. However, this is not unlike more laborious approaches to transducer design which we have at present. Quite often, similar forms of design process can be applied in the development of a transducer concept, but inevitably each class or type of transducer possesses key design elements that do not directly translate. For example, the operating modes of a flexural ultrasonic transducer are effectively dependent on standard plate modes of vibration, as identified through Leissa's theory of plate vibrations (Leissa, 1969). The optimisation of a flexural ultrasonic transducer using this plate vibration theory for a transducer operating in ambient conditions of pressure and temperature in air is a technically different problem, and in some ways more complex, to that where pressure and temperature fluctuations must be considered. Yüksel et al. also note the requirement for sufficient training data (Yüksel et al., 2023), and of the required quality. There are several areas of ultrasound research where that quantity of data is still underdeveloped, but with progress in the field in general, there will be new opportunities presented. There is little doubt that with the advent of more advanced artificial intelligence techniques and more effective ways of administering them, there is the possibility that a wide range of advanced adaptive ultrasonic devices can be realised, and at a faster rate than previously thought possible.

Many components used in the assembly of a transducer could be optimised using artificial intelligence techniques including artificial neural networks, such as the shape and material properties of a Langevin transducer end-mass to which an end-effector is attached, and the geometry of the device for a particular operating mode of interest. There could also be adaptive techniques for the control and positioning of the end-effector, to deliver a device response tailored to a specific operation. As just one example, there could be an intricate dental procedure required, necessitating a high level of precision for a cutting blade. Artificial intelligence techniques could potentially be used in such a circumstance to locate the cutting tip of the ultrasonic device in the position that would be optimal for patient outcomes. Example input data for this process may be those related to human factors, including the condition of the patient's mouth and its geometry, the nature of the cutting tip and its physical parameters, the type of operation required, including the option of diagnosis, and the intended outcome. Of course, this is not an exhaustive list, and even this set of input data alone may potentially require many hours of training, depending on the computational facilities available. The strategies for incorporating intelligence into ultrasonic devices in the pursuit of highly advanced, adaptive devices will likely take significant time and effort. The procurement of the highest quality data, in the quantities we need for the reliable engineering of a new generation of ultrasound technologies, is an important first step.

5.5 SUMMARY

If we return to one of the core underpinning principles of this chapter, that is, the pathway to ultrasonic technologies we may consider to be truly intelligent, then we must consider how intelligence can be incorporated into ultrasonic technologies. The key advances in adaptive ultrasonics concerning transducer design and the selection of advanced or smart materials were presented across Chapters 3 and 4. Whilst smart materials such as shape memory alloys undoubtedly present opportunities for adaptive ultrasonics, they do not yet present the capabilities to introduce intelligent, decision-making properties into an ultrasonic device. One of the key advances in the forthcoming generations of ultrasound technology will likely be the intelligent selection of ultrasonic devices and procedures for a specific application. This would have the potential to revolutionise the efficacy of a wide range of ultrasonic processes, processing time, and perhaps significantly improve patient outcomes, where relevant.

In general, this chapter has documented the progress in the concept of adaptive ultrasonics via the opportunities afforded by smart signal processing strategies, artificial intelligence techniques, and adaptive beamforming which have all become popular in medical and industrial ultrasound imaging. Further to this, a brief overview of the opportunities in the optimisation of the engineering design process has been provided, with discussion from the point of view of artificial intelligence, outlining how such approaches can potentially be used in a responsible way to optimise the design features of different classes of ultrasonic transducer. Given where the technology landscape currently is, and with ultrasound imaging and beamforming techniques dominating advances in adaptive ultrasonic technologies, there remains significant progress to be made to realise a new generation of adaptive ultrasonic technologies for the applications presented throughout this book, from power ultrasonics including cavitation and cutting or drilling, to those which are nondestructive in their primary function such as sensing and imaging. The proliferation of data across all aspects of ultrasound research will drive adaptive technologies forward, in all the key areas which have been discussed thus far in this book. The next steps in development should consider elements of modern practice and how these technologies will likely mature over the forthcoming years.

6 Modern Practice

6.1 PROJECTIONS FOR THE FUTURE

It is difficult to accurately predict the future development of adaptive ultrasonics, but it is evident that forthcoming innovations will be centred around three major areas. The first is in the continued development of **advanced manufacturing** techniques. The manufacturing methods outlined in this book, for example, the electrical discharge machining approach of fabricating transducer components composed of shape memory alloys like Nitinol, or the range of additive manufacturing methods available for synthesising a wide array of metamaterials and metastructures, will continue to be innovated and exploited over the coming decades, because they have been demonstrated to be reliable and effective in the fabrication process. The details provided in this book have illustrated that whilst effective and key to the development of shape memory alloy-based adaptive ultrasonic transducers, a technique such as electrical discharge machining can be prohibitively slow, and the uptake of these devices may take a long period of time since they remain relatively expensive and time consuming to produce. With continued investment, research and development, and impact-oriented case studies with key industries, the benefits of the technology will progressively emerge. From this, new applications will come to the forefront as technology develops.

There are continued developments in additive manufacturing, both in terms of metallic materials and for resins more commonly utilised in metamaterial research. In the contexts of acoustics and ultrasonics, the additive manufacturing of metals for devices remains in a relative stage of infancy, with fundamental studies available of integrating additively manufactured metals into transducers. However, much further research and development is required because there is not yet a complete picture of how additive manufacturing can be used to deliver high-performance ultrasonic devices, particularly for higher power applications, exhibiting the necessary mechanical resilience and longevity displayed by conventional transducers. It is possible to use additive manufacturing techniques to print shape memory alloys, and this book has outlined some elementary developments in how this has been achieved for Nitinol. In the contexts of ultrasonic transducers and their adaptive dynamic properties, a programme of research will be required to assess their operational capabilities and to understand the mechanisms by which their transformational properties can be tailored and realised. There are continuing efforts to enhance the fatigue life properties of additively manufactured metallic components, and it would be expected that a similar drive would be undertaken for additively manufactured shape memory materials.

The other relevant aspect of advanced manufacturing will be the continued development of a range of shape memory materials which can compete or perhaps

DOI: 10.1201/9781003324126-6

replace Nitinol in adaptive ultrasonic transducers. The benefits of adopting Nitinol as the shape memory material of interest in current device configurations are that it is a readily available material, it has a deep repository of research associated with it which can be traced back in decades, and it is relatively straightforward to manufacture and optimise compared to some more intricate or complex compositions. It also tends to be able to be tailored to operate at temperatures conducive to practical engineering applications, such as for the biomedical industry or as simple actuators for operation at ambient room temperature. A broader spectrum of applications will emerge, and so there will be a necessity to innovate and engineer a series of shape memory alloys able to deliver the transformational properties required. For example, Nitinol would not be well suited to hot environments above 100°C, and so alternative candidates of alloys would need to be sought in those circumstances. There is clearly the scope to achieve this through the continued development of materials science, with a view to optimising the integration of these materials into devices.

Up to this point, adaptive ultrasonic transducers have tended to predominantly exploit the mechanical characteristics of the shape memory materials, notably the elastic or Young's modulus. The major reason for this is that the resonant behaviour of an acoustic or ultrasonic transducer is highly dependent on this material property. Future innovation pathways will investigate the transformational influence of multiple material properties on the performance of adaptive ultrasonic devices. For example, and which has been outlined in this book and prominently in Chapter 4, there are several material properties of Nitinol which are dependent on the phase microstructure of the material. One of these is the electrical resistivity of the material, for instance. It would therefore not be unreasonable to suggest that a broad range of phase-sensitive properties could emerge from a series of shape memory materials, that are controllable, and which would introduce several adaptive features into an acoustic or ultrasonic device which could be harnessed simultaneously. From here, there would be the potential to engineer truly innovative and complex multi-parameter transformational adaptive ultrasonic devices.

The second major area of future development into the 21st century will be centred around our capacity to deliver more precise and powerful **computational processing** techniques. One side of this can be regarded as an extension of what is contained in Chapter 5, with more advanced protocols forthcoming to handle more demanding applications, to rapidly accelerate the responsiveness of devices, and those which can tailor the device response to the changing conditions of an environment. It would be expected that such computational approaches would be implemented in concepts making use of metamaterials. As has been discussed in Chapter 5, metamaterials and metastructures provide the ideal platform for promoting controllable beam shaping or beamforming in an ultrasonic system. It may be possible to engineer such systems to generate tailored wave propagation patterns using bespoke arrays which can be readily additively manufactured, or even through a metastructure-like configuration which transforms the approach considering the traditional limitations of engineering multi-element arrays. There would then be the potential to revolutionise how energy conversion in such systems can be optimised in the future, and perhaps utilised to undertake other functions as part of a larger system.

The third projection is the continued progression towards **miniaturisation** and the realisation of more portable and adaptable configurations of ultrasonic systems. This is not limited to adaptive ultrasonics, because current progress is already driving conventional or more popular applications of ultrasonic technology towards miniaturisation. For example, there have been significant developments in miniaturised devices for ultrasonic surgery. Some of these developments have required the use of innovative approaches to manufacturing, a few of which have been outlined thus far in this book. The additive manufacturing of metals to realise complex device structures which cannot be delivered using conventional machining techniques will continue to play a significant role in how ultrasonic devices can be miniaturised, and this also extends to those fabricated using shape memory materials. The transition towards greater miniaturisation will place additional challenges on how advanced materials such as shape memory alloys can be integrated with acoustic and ultrasonic transducers. For example, far greater levels of precision will be required in the design and manufacture of transducer components, and this will require continued development in additive manufacturing technologies as well as precision in machining, through diamond-assisted or electrical discharge approaches. It is also likely that laser-assisted machining methods may be considered more widely for these materials, with closer collaboration with material manufacturers to synthesise and fabricate bespoke material configurations.

6.2 ROUTES TO SUCCESS

There are many ways in which technology innovations can come about, but little can happen without the proper investment. In particular, the fabrication and processing of shape memory materials and metamaterials can be time consuming and expensive, and the necessary developments to realise the next generation of adaptive ultrasonic and acoustic devices will not be possible without this. It will not happen in the short term, but the costs associated with the design and fabrication of these materials must progressively reduce for them to receive the uptake in a broad range of medical and industrial applications they will need to be successful. There is arguably faster progress possible on the metamaterials front, because of the many ways they can be rapidly additively manufactured, and the resins available to do this. The principal challenge for metamaterials is how the underlying mathematics can be handled to deliver the optimised configurations necessary for adaptive transducer concepts.

This leads on to a particularly important aspect of how success can be achieved in the discipline of adaptive ultrasonics. It is a bridge between complex materials science and the field of ultrasonics, requiring understanding of electronics, mechanics, dynamics, and mathematics, to create viable products. Materials science and ultrasonics are traditionally disparate disciplines, but like many areas of engineering and science, there are emerging interfaces of activity. It will be important for a new generation of engineers to understand how to bridge these different facets of science together to innovate a new generation of technology. Due to the expense associated with shape memory materials and metamaterial manufacture, they are resources which are not always available to students or even recent graduates, and so there can be restrictions in learning and who is able to become proficient in the underlying science. One way in which this can be addressed is through much more open collaboration between different academic research groups, key industries, and interested manufacturers of materials. There is already positive progress in this regard, and the literature fortunately contains many valuable articles of interest which are open access. Some instructive practical resources would also be important in helping scientists and engineers at earlier stages in their careers to become more familiar with these technologies and their potential. It would also be important to try and exploit the capabilities we now have in computational processing, particularly artificial intelligence approaches, to optimise the design and implementation of adaptive ultrasonic devices. This has been addressed in some way in Chapter 5, but it is worth reinforcing the point that experts in many fields have a role to play in the progression and innovation of this technology.

It is also possible that regulatory changes will drive us to engineer new configurations of ultrasonic transducer. The continued review of the Restriction of Hazardous Substances (RoHS) associated with using the element lead in piezoelectric ceramic materials means that for the time being, there is an element of *business as usual* for the design and manufacture of ultrasonic transducers. However, in the event of the conditions around this exemption changing, it may force some changes into how ultrasonic transducers are designed and manufactured. The materials and methods outlined in this book then may play a role in how future ultrasonic technologies can be designed and optimised. There is a whole array of research investigating high-performance piezoelectric materials which do not require the presence of lead in their compositions, and so it is likely that a future route to success will bridge this research with that associated with the advanced materials reported in this book.

There should certainly be closer collaboration between device designers, the manufacturers of the constituent materials (most importantly the active driver materials such as the piezoelectric ceramics and those responsible for synthesising the advanced materials, whether they are shape memory or metamaterials), and the end users. In future, there will continue to be a greater focus on all aspects of a product lifecycle which can be viewed holistically, making use of the different phases of a product's life to optimise performance or application. As an illustration, the nature of shape memory materials means that they are particularly well suited to the approach of *design for remanufacturing*. In general, many compositions constitute extremely mechanically resilient materials, and it is likely that an embedded piezoelectric ceramic material will crack, depolarise, or degrade, before a shape memory material like Nitinol will degrade or lose its transformational properties. It would theoretically be possible to extract a shape memory alloy such as Nitinol from an existing adaptive ultrasonic transducer and repurpose it to be integrated into another device, at the 'end of life' of its current configuration. There would hence be some interesting and environmentally sustainable approaches to device design which could emerge, as part of a growing transition to more sustainable and resource-efficient economies in the future. This is extremely important, because of the difficulty of manufacturing such shape memory alloy components and the costs associated with their initial fabrication. One consideration would be how to qualify a component such as this for reuse in another device. There would need to be some form of assessment protocol put in place, to gain confidence in its safe and effective use in a subsequent device or system.

As a device designer, it can be easy to overlook the importance of close communication and collaboration with the manufacturers of the materials that are being used to design and construct the transducers. As constituent materials of transducers become more complex, there is a greater need for ensuring close collaborations, and this helps manufacturers understand more about their own materials. A principal challenge for a material such as Nitinol relates to measuring and understanding its modulus properties, and it would be expected that future innovations in devices would uncover a range of effective strategies to address this. Similarly, a success criterion for adaptive ultrasonic transducer design can be in the consideration of the end user of the device, noting the requirements of the system and how it will be operated in practice. Undertaking this early in the transducer design process can eliminate many of the practical problems which can emerge later, such as fixture or constraint requirements, or changes in transducer dynamics associated with variations in environmental temperature associated with the target application.

6.3 SUMMARY AND OUTLOOK

There is much cause to be optimistic about the trajectory of advanced manufacturing and our computational capabilities. Already there have been rapid developments in adaptive ultrasonic devices since 2000, before which point there was comparatively little regarding practical metamaterials demonstrated, and Newnham and Meyer, Jr.'s research into a Nitinol based flextensional ultrasonic transducer had only just been published in the *Journal of Intelligent Material Systems and Structures*. In the near quarter of a century since then, there has been a rapid acceleration of shape memory material and metamaterial research, including our abilities to synthesise and fabricate a wide range of configurations and compositions. At the same time, there have been arguably faster developments in computational power and the resolution at which ultrasonic measurements can be made. Machine learning techniques and artificial intelligence approaches are now being integrated into the field of ultrasonics, and there does not appear to be any slowing of this trend. It would be beneficial for such computational facilities to be employed in each phase of the lifecycle of an adaptive ultrasonic device, including in the design phase. Advanced materials, in general, are complex, and as such the devices from which they are manufactured are highly complicated in their nature, both in terms of the electromechanical response but also their transformational behaviours. Artificial intelligence techniques could therefore prove highly effective in how such devices are optimally designed and manufactured in future, accounting for the myriad of adaptive properties they incorporate.

It is likely that the research and development landscape throughout the 21st century will continue to prominently feature elements of intelligent device and system design and sustainable engineering, and this is where adaptive ultrasonics can play a significant role. The tuneable and transformational behaviours associated with these devices have the potential to enable truly versatile and intelligent devices and open possibilities to create systems with greater longevity and opportunities to replace or remove materials from conventional configurations of device which are not consistent with the future vision of a sustainable engineering landscape.

Appendix A: An Analytical Model of Ultrasonic Device Response

The complete derivation of the vibration response of a transducer is provided here and directly extracted from the open access material, via a CC-BY 4.0 licence, published by the author in 2018, regarding the investigation of the dynamic response of a flexural ultrasonic transducer (Feeney et al., 2018a). Despite the original focus of this research on the dynamic performance of the flexural ultrasonic transducer, the core principle can logically be extended to a variety of ultrasonic transducers, including those which have been covered in this book. The modelling approach should hopefully serve as a useful tool and starting point for the reader. Note that the key steps in this analytical model are defined using bracketed numbers in the form (n), from (A.1) through to (A.33).

First, the dynamic response of the transducer can be expressed as a combination of mass M, stiffness K, and damping C terms, with a forcing function F convolved with an appropriate signal to enable us to accurately represent the build-up of vibration motion from around zero to a steady-state condition.

$$M\ddot{x} + C\dot{x} + Kx = F\sin(\omega t).H(t_0 - t) \tag{A.1}$$

In (A.1), the t_0 parameter is the time that the forcing function is ended, with $H(t)$ being the Heaviside function. In this relationship, both x and $\dot{x}$ are zero at a time of $t = 0$. Therefore, a solution for (A.1) exists for the case of $t < t_0$, and this can be demonstrated through (A.2), followed by the associated derivatives, of the first and the second order, respectively, through (A.3) and (A.4).

$$x = F_+ e^{\lambda_+ t} + F_- e^{\lambda_- t} + A\sin(\omega t) + B\cos(\omega t) \tag{A.2}$$

$$\dot{x} = \lambda_+ F_+ e^{\lambda_+ t} + \lambda_- F_- e^{\lambda_- t} + A\omega\cos(\omega t) - B\omega\sin(\omega t) \tag{A.3}$$

$$\ddot{x} = \lambda_+^2 F_+ e^{\lambda_+ t} + \lambda_-^2 F_- e^{\lambda_- t} - A\omega^2\sin(\omega t) - B\omega^2\cos(\omega t) \tag{A.4}$$

In (A.2), the F_+ and F_- terms can be regarded as homogeneous, and therefore can be used to simulate the natural resonance of a transducer. In contrast, the $A\sin$ and $B\cos$ parameters both relate to the forcing on the system, or in other words, the excitation condition. This could be an applied driving voltage to the transducer. As an extension to this system of equations, (A.1) can be used to generate a general solution for $H = 1$ when $0 < t \leq t_0$. This is exhibited by (A.5).

$$M\ddot{x} + C\dot{x} + Kx = F\sin(\omega t) \tag{A.5}$$

If we compare (A.2), (A.3), and (A.4) with (A.5), and if we gather the exponential entries in the equations together, then we can produce (A.6). This formula can then be solved to generate relationships for λ_+ and λ_-, as depicted in (A.7). Then, by gathering the $\sin(\omega t)$ and $\cos(\omega t)$ quantities together, we can find relationships for A and B, as provided in (A.8) through to (A.13).

$$M\lambda_{\pm}^2 + C\lambda_{\pm} + K = 0 \tag{A.6}$$

$$\lambda_{\pm} = \frac{-C \pm \sqrt{C^2 - 4MK}}{2M} \tag{A.7}$$

$$B\left(-M\omega^2 + K\right) + C\omega A = 0 \tag{A.8}$$

$$A\left(-M\omega^2 + K\right) - C\omega B = F \tag{A.9}$$

$$B = \frac{C\omega A}{M\omega^2 - K} \tag{A.10}$$

$$A\left(-M\omega^2 + K\right) - C\omega\left(\frac{C\omega A}{M\omega^2 - K}\right) = F \tag{A.11}$$

$$A\left(-M\omega^2 + K\right)\left(M\omega^2 - K\right) - C^2\omega^2 A = F\left(M\omega^2 - K\right) \tag{A.12}$$

$$A = \frac{F\left(M\omega^2 - K\right)}{\left(-M\omega^2 + K\right)\left(M\omega^2 - K\right) - C^2\omega^2} = gF \tag{A.13}$$

From here, the amplitudes of the natural resonances can be determined, and therefore through the application of the defined initial conditions that both x and $\dot{x}$ are zero at a time of $t = 0$ for (A.2) and (A.3), then we can generate (A.14) through to (A.19).

$$F_+ + F_- + B = 0 \tag{A.14}$$

$$F_- = -B - F_+ \tag{A.15}$$

$$\lambda_+ F_+ + \lambda_- F_- + A\omega = 0 \tag{A.16}$$

$$\lambda_+ F_+ = -\lambda_- F_- - A\omega \tag{A.17}$$

$$\lambda_+ F_+ = \lambda_- B + \lambda_- F_+ - A\omega \tag{A.18}$$

$$F_+\left(\lambda_+ - \lambda_-\right) = \lambda_- B - A\omega \tag{A.19}$$

If the expression shown in (A.7) is now utilised, both λ_+ and λ_- can be used to formulate the relationships shown by (A.20) and (A.21). From here, (A.22) can be expressed allowing the F_+ and F_- parameters in (A.23) and (A.24) to be determined. These expressions can be directly applied in the general formula for the vibration condition depending on the damping level.

$$\lambda_+ = \frac{-C + \sqrt{C^2 - 4MK}}{2M} \tag{A.20}$$

$$\lambda_- = \frac{-C - \sqrt{C^2 - 4MK}}{2M} \tag{A.21}$$

$$\left(\lambda_+ - \lambda_-\right) = \frac{\sqrt{C^2 - 4MK}}{2M} \tag{A.22}$$

$$F_+ = \frac{B\left(\dfrac{-C - \sqrt{C^2 - 4MK}}{2M}\right) - A\omega}{\left(\dfrac{\sqrt{C^2 - 4MK}}{2M}\right)} \tag{A.23}$$

$$F_- = \frac{\left(\dfrac{-C + \sqrt{C^2 - 4MK}}{2M}\right)F_+ + A\omega}{-\left(\dfrac{C - \sqrt{C^2 - 4MK}}{2M}\right)} \tag{A.24}$$

If an ultrasonic transducer is to be assumed to be operating in a state of resonance and hence in an underdamped condition, as may be common, then an associated set of solutions can be generated for the equation of motion. These are provided in (A.25)–(A.27).

$$x(t) = F_+ e^{(-\alpha + i\bar{\alpha})t} + F_- e^{(-\alpha - i\bar{\alpha})t} + A\sin(\omega t) + B\cos(\omega t) \tag{A.25}$$

$$x(t) = e^{-\alpha t}e^{i\alpha t}F_+ + e^{-\alpha t}e^{-i\alpha t}F_- + A\sin(\omega t) + B\cos(\omega t) \tag{A.26}$$

$$x(t) = F_+ e^{-\alpha t}\left\{\left(\cos\bar{\alpha}t\right) + \left(i\sin\bar{\alpha}t\right)\right\} +$$

$$F_- e^{-\alpha t}\left\{\left(\cos\bar{\alpha}t\right) - \left(i\sin\bar{\alpha}t\right)\right\} + A\sin(\omega t) + B\cos(\omega t) \tag{A.27}$$

The A and B parameters in (A.25)–(A.27) are directly related to the forced excitation on the transducer, indicative of the amplitude condition. Prior to the vibration response of the transducer reaching steady-state, it will change with time depending on the proximity of the excitation frequency from resonance, and the time taken for the amplitude response to build up to the steady-state condition. It is common to designate the amplitude response of the transducer in this case, by a term such as $\bar{E}$, which is shown in the expression indicated by (A.32). The preceding derivation process, at a fundamental level and shown in (A.28) through to (A.31), can be built up by considering the solution to (A.1). This amplitude term $\bar{E}$ is effectively the square root of the sum of the squares of A and B. It is also important to incorporate the phase parameter Φ in the formulation, allowing a comprehensive derivation of the vibration response of a transducer in the time domain, for a given drive frequency and amplitude and accounting for the resonance frequency, which may differ from that of the drive. In this way, the derived expressions for B and A shown via (A.10) and (A.13), respectively, can be used to generate a concise and effective relationship for $\bar{E}$, detailed in (A.32).

$$x(t) = F_{+}e^{-\alpha t}\left(\cos\bar{\alpha}t + i\sin\bar{\alpha}t\right) + F_{-}e^{-\alpha t}\left(\cos\bar{\alpha}t - i\sin\bar{\alpha}t\right)$$

$$+\sqrt{A^2 + B^2}\left(\sin(\omega t + \Phi)\right) \text{ where } \tan\Phi = \frac{B}{A} \tag{A.28}$$

$$A^2 + B^2 = A^2 + A^2\left(\frac{C\omega}{M\omega^2 - K}\right)^2 \tag{A.29}$$

$$A^2 + B^2 = (gF)^2\left\{1 + \left(\frac{C\omega}{M\omega^2 - K}\right)^2\right\} \tag{A.30}$$

$$A^2 + B^2 = (gF)^2\left(\frac{\left(M\omega^2 - K\right)^2 + \left(C\omega\right)^2}{\left(M\omega^2 - K\right)^2}\right) \tag{A.31}$$

$$\bar{E} = \sqrt{A^2 + B^2} = \frac{gF}{\left(M\omega^2 - K\right)}\sqrt{\left(M\omega^2 - K\right)^2 + \left(C\omega\right)^2} \tag{A.32}$$

The parameters generated in this section can all be used to obtain an amplitude spectrum for a given set of initial or assumed conditions, for a particular operation. The specific vibration response of a transducer can be determined via (A.33) accounting for the natural resonance (F) and the drive (E) parts together.

$$x(t) = (F_{+} + F_{-})\left(e^{-\alpha t}\cos\bar{\alpha}t\right) + \bar{E}\left(\sin(\omega t + \Phi)\right) \tag{A.33}$$

Readers are encouraged to trial the solution given here for various use cases relating to ultrasonic transducers and their applications, for example, as a rapid indicator of operational performance for a range of excitation conditions. Selections of results can also be found in the literature, particularly related to that of the flexural ultrasonic transducer, for example, in Dixon et al. (2017) and Feeney et al. (2017b, 2018a).

Appendix B: Selected Multimedia Resources

The following links are useful resources for further information relating to the content of this book, each of which the reader is invited to review. It should be noted that the external links to the content were correct at the time of publication.

Theme	Link
Modal Analysis through FEA	https://www.youtube.com/watch?v=AIMAqBatD-k
Flexural Ultrasonic Transducers with Mode Frequency Calculator	https://warwick.ac.uk/fac/sci/physics/research/ultra/research/hiffut/
MathWorks® Signal Processing	https://www.mathworks.com/help/dsp/
Fresnel and Fraunhofer Calculator	https://www.everythingrf.com/rf-calculators/antenna-near-field-distance-calculator
Shape Memory Alloys	https://www.stem.org.uk/resources/community/collection/449829/shape-memory-alloys
TED-Ed on Shape Memory Alloys	https://www.youtube.com/watch?v=yR-6_lS9vts

FIGURE A2.1 Summary table with details of multimedia resources of interest.

UKAN Design of Acoustic Metamaterials	https://www.youtube.com/watch?v=aqasijejwas
Machine Learning for Ultrasound Medical Imaging	https://www.youtube.com/watch?v=RECqC2HqtJU
Practical Aspects of Ultrasonic Transducer Design	https://youtube.com/watch?v=YwPeHK3aJOU
Harmonic Analysis through FEA	https://www.youtube.com/watch?v=qwFuXR-R8OI

FIGURE A2.1 (CONTINUED) Summary table with details of multimedia resources of interest.

Bibliography

Adams, T.M., Kirkpatrick, S.R., Wang, Z. and Siahmakoun, A., 2005. NiTi shape memory alloy thin films deposited by co-evaporation. *Materials Letters*, 59(10), pp. 1161–1164.

Adharapurapu, R.R., 2007. *Phase transformations in nickel-rich nickel-titanium alloys: Influence of strain-rate, temperature, thermomechanical treatment and nickel composition on the shape memory and superelastic characteristics.* University of California, San Diego.

Afzal, M.S., Shim, H. and Roh, Y., 2018. Design of a piezoelectric multilayered structure for ultrasound sensors using the equivalent circuit method. *Sensors*, 18(12), p. 4491.

Agarwal, N., Bourke, D., Obeidi, M.A. and Brabazon, D., 2024. Influence of laser powder bed fusion and ageing heat treatment parameters on the phase structure and physical behaviour of Ni-rich nitinol parts. *Journal of Materials Research and Technology*, 30, pp. 4527–4541.

Akl, W. and Baz, A., 2012. Analysis and experimental demonstration of an active acoustic metamaterial cell. *Journal of Applied Physics*, 111(4), p. 044505.

Al Mukaddim, R., Ahmed, R. and Varghese, T., 2021. Subaperture processing-based adaptive beamforming for photoacoustic imaging. *IEEE Transactions on Ultrasonics, Ferroelectrics, and Frequency Control*, 68(7), pp. 2336–2350.

Al-Budairi, H., Harkness, P. and Lucas, M., 2011. A strategy for delivering high torsionality in longitudinal-torsional ultrasonic devices. *Applied Mechanics and Materials*, 70, pp. 339–344.

Al-Budairi, H., Lucas, M. and Harkness, P., 2013. A design approach for longitudinal–torsional ultrasonic transducers. *Sensors and Actuators A: Physical*, 198, pp. 99–106.

Alapati, S.B., Brantley, W.A., Iijima, M., Clark, W.A., Kovarik, L., Buie, C., Liu, J. and Johnson, W.B., 2009. Metallurgical characterization of a new nickel-titanium wire for rotary endodontic instruments. *Journal of Endodontics*, 35(11), pp. 1589–1593.

Alipour, S., Taromian, F., Ghomi, E.R., Zare, M., Singh, S. and Ramakrishna, S., 2022. Nitinol: From historical milestones to functional properties and biomedical applications. *Proceedings of the Institution of Mechanical Engineers, Part H: Journal of Engineering in Medicine*, 236(11), pp. 1595–1612.

Allam, A., Elsabbagh, A. and Akl, W., 2016. Modeling and design of two-dimensional membrane-type active acoustic metamaterials with tunable anisotropic density. *The Journal of the Acoustical Society of America*, 140(5), pp. 3607–3618.

Alobaidi, W.M., Alkuam, E.A., Al-Rizzo, H.M. and Sandgren, E., 2015. Applications of ultrasonic techniques in oil and gas pipeline industries: A review. *American Journal of Operations Research*, 5(04), p. 274.

Alwi, H.A., Smith, B.V. and Carey, J.R., 1996. Factors which determine the tunable frequency range of tunable transducers. *The Journal of the Acoustical Society of America*, 100(2), pp. 840–847.

Alzhanov, N., Tariq, H., Amrin, A., Zhang, D. and Spitas, C., 2023. Modelling and simulation of a novel nitinol-aluminium composite beam to achieve high damping capacity. *Materials Today Communications*, 35, p. 105679.

Amaral, J.F. and Chrostek, C.A., 1997. Experimental comparison of the ultrasonically-activated scalpel to electrosurgery and laser surgery for laparoscopic use. *Minimally Invasive Therapy & Allied Technologies*, 6(4), pp. 324–331.

Antonucci, V., Faiella, G., Giordano, M., Mennella, F. and Nicolais, L., 2007. Electrical resistivity study and characterization during NiTi phase transformations. *Thermochimica Acta*, 462(1–2), pp. 64–69.

Ashurst, G.R., 2002. Ambient temperature shape memory alloy actuator. US Patent 6,427,712 B1.

ASTM F2004-05, 2004. Standard test method for transformation temperature of nickel-titanium alloys by thermal analysis. *ASTM Standard*, 5, pp. 1–4.

ASTM F2516-07, 2007. Standard test method for tension testing of nickel-titanium superelastic materials.

ASTM F2082, 2016. Standard test method for determination of transformation temperature of nickel-titanium shape memory alloys by bend and free recovery.

ASTM F2063-18, 2018. Standard specification for wrought nickel-titanium shape memory alloys for medical devices and surgical implants.

Athanassiadis, A.G., Ma, Z., Moreno-Gomez, N., Melde, K., Choi, E., Goyal, R. and Fischer, P., 2021. Ultrasound-responsive systems as components for smart materials. *Chemical Reviews*, 122(5), pp. 5165–5208.

Bach, C., Kabir, M.N., Goyal, A., Malliwal, R., Kachrilas, S., Howairis, M.E.E., Masood, J., Buchholz, N. and Junaid, I., 2013. A self-expanding thermolabile nitinol stent as a minimally invasive treatment alternative for ureteral strictures in renal transplant patients. *Journal of Endourology*, 27(12), pp. 1543–1545.

Bale, A.S., Reddy, S.V. and Tiwari, S., 2020, June. Effect of residual stress on resonant frequency in Nitinol based thin film resonator. In *IOP Conference Series: Materials Science and Engineering* (Vol. 872, No. 1, p. 012008). IOP Publishing.

Bejarano, F., Feeney, A. and Lucas, M., 2014. A cymbal transducer for power ultrasonics applications. *Sensors and Actuators A: Physical*, 210, pp. 182–189.

Bejarano, F., Feeney, A., Wallace, R., Simpson, H. and Lucas, M., 2016. An ultrasonic orthopaedic surgical device based on a cymbal transducer. *Ultrasonics*, 72, pp. 24–33.

Bell, A.J. and Deubzer, O., 2018. Lead-free piezoelectrics—The environmental and regulatory issues. *MRS Bulletin*, 43(8), pp. 581–587.

Belyaev, S., Volkov, A. and Resnina, N., 2014. Alternate stresses and temperature variation as factors of influence of ultrasonic vibration on mechanical and functional properties of shape memory alloys. *Ultrasonics*, 54(1), pp. 84–89.

Blitz, J., 1963. *Fundamentals of Ultrasonics*. Butterworth & Co. (Publishers) Ltd.

Boota, M., Houwman, E.P., Lanzara, G. and Rijnders, G., 2023. Effect of a niobium-doped PZT interfacial layer thickness on the properties of epitaxial PMN-PT thin films. *Journal of Applied Physics*, 133(14), p. 145302.

Brammajyosula, R., Buravalla, V. and Khandelwal, A., 2011. Model for resistance evolution in shape memory alloys including R-phase. *Smart Materials and Structures*, 20(3), p. 035015.

Caleon, I.S. and Subramaniam, R., 2007. From Pythagoras to Sauveur: Tracing the history of ideas about the nature of sound. *Physics Education*, 42(2), p. 173.

Cambridge Dictionary Online. n.d. https://dictionary.cambridge.org/dictionary/english/adaptive

Cheeke, J.D.N., 2010. *Fundamentals and Applications of Ultrasonic Waves*. CRC Press.

Chen, J., Xu, Q.C., Blaszkiewicz, M., Meyer Jr, R. and Newnham, R.E., 1992. Lead zirconate titanate films on nickel–titanium shape memory alloys: SMARTIES. *Journal of the American Ceramic Society*, 75(10), pp. 2891–2892.

Chen, J., Li, Z. and Zhao, Y.Y., 2009. A high-working-temperature CuAlMnZr shape memory alloy. *Journal of Alloys and Compounds*, 480(2), pp. 481–484.

Chen, K., Yao, A., Zheng, E.E., Lin, J. and Zheng, Y., 2012. Shear wave dispersion ultrasound vibrometry based on a different mechanical model for soft tissue characterization. *Journal of Ultrasound in Medicine*, 31(12), pp. 2001–2011.

Chen, S., Fan, Y., Fu, Q., Wu, H., Jin, Y., Zheng, J. and Zhang, F., 2018. A review of tunable acoustic metamaterials. *Applied Sciences*, 8(9), p. 1480.

Cheng, B.L., Gabbay, M., Duffy, W. and Fantozzi, G., 1996. Mechanical loss and Young's modulus associated with phase transitions in barium titanate based ceramics. *Journal of Materials Science*, 31, pp. 4951–4955.

Chipana, A., Hill, J.R. and Dillon, C.R., 2023, September. In situ actuation of shape memory alloy using focused ultrasound. In *Smart Materials, Adaptive Structures and Intelligent Systems* (Vol. 87523, p. V001T04A005). American Society of Mechanical Engineers.

Cleary, R., Wallace, R., Simpson, H., Kontorinis, G. and Lucas, M., 2022. A longitudinal-torsional mode ultrasonic needle for deep penetration into bone. *Ultrasonics*, 124, p. 106756.

Costanza, G. and Tata, M.E., 2020. Shape memory alloys for aerospace, recent developments, and new applications: A short review. *Materials*, 13(8), p. 1856.

Cracknell, A.P., 1980. *Ultrasonics*. Wykeham Publications.

Craig, S.R., Welch, P.J. and Shi, C., 2020. Non-Hermitian complementary acoustic metamaterials for imaging through skull with imperfections. *Frontiers in Mechanical Engineering*, 6, p. 55.

Cummer, S.A., Christensen, J. and Alù, A., 2016. Controlling sound with acoustic metamaterials. *Nature Reviews Materials*, 1(3), pp. 1–13.

Dahan, E. and Cohen, I., 2023. Deep-learning-based multitask ultrasound beamforming. *Information*, 14(10), p. 582.

Datla, N.V., Honarvar, M., Nguyen, T.M., Konh, B., Darvish, K., Yu, Y., Dicker, A.P., Podder, T.K. and Hutapea, P., 2012, September. Towards a nitinol actuator for an active surgical needle. In *Smart Materials, Adaptive Structures and Intelligent Systems* (Vol. 45103, pp. 265–269). American Society of Mechanical Engineers.

David, G., Robert, J.L., Zhang, B. and Laine, A.F., 2015. Time domain compressive beam forming of ultrasound signals. *The Journal of the Acoustical Society of America*, 137(5), pp. 2773–2784.

DeAngelis, D.A., Schulze, G.W. and Wong, K.S., 2015. Optimizing piezoelectric stack preload bolts in ultrasonic transducers. *Physics Procedia*, 63, pp. 11–20.

Demi, L., 2018. Practical guide to ultrasound beam forming: Beam pattern and image reconstruction analysis. *Applied Sciences*, 8(9), p. 1544.

Dixon, S., Kang, L., Ginestier, M., Wells, C., Rowlands, G. and Feeney, A., 2017. The electro-mechanical behaviour of flexural ultrasonic transducers. *Applied Physics Letters*, 110(22), 223502.

Donald, I. and Abdulla, U., 1967. Further advances in ultrasonic diagnosis. *Ultrasonics*, 5(1), pp. 8–12.

Dixon, S., Kang, L., Feeney, A. and Somerset, W.E., 2021. Active damping of ultrasonic receiving sensors through engineered pressure waves. *Journal of Physics D: Applied Physics*, 54(13), p. 13LT01.

Dogra, S. and Gupta, A., 2021, October. Design, manufacturing, and acoustical analysis of a Helmholtz resonator-based metamaterial plate. In *Acoustics* (Vol. 3, No. 4, pp. 630–641). MDPI.

Donald, I., 1974. Sonar—The story of an experiment. *Ultrasound in Medicine & Biology*, 1(2), pp. 109–117.

Duck, F.A. and Thomas, A., 2022. Paul Langevin (1872–1946): The father of ultrasonics. *Medical Physics International Journal*, 10(1), pp. 84–91.

Duerig, T.W. and Pelton, A.R., 1994. Ti-Ni shape memory alloys. In *Materials Properties Handbook: Titanium Alloys*, 1 (pp. 1035–1048). https://scholar.google.com/citations? view_op=view_citation&hl=en&user=QPEYYhEAAAAJ&citation_for_view= QPEYYhEAAAAJ:pyW8ca7W8N0C

Duerig, T., Pelton, A. and Stöckel, D.J.M.S., 1999. An overview of nitinol medical applications. *Materials Science and Engineering: A*, 273, pp. 149–160.

Duerig, T., Stoeckel, D. and Johnson, D., 2003, March. SMA: Smart materials for medical applications. In *European Workshop on Smart Structures in Engineering and Technology* (Vol. 4763, pp. 7–15). SPIE.

Duerig, T.W., 2006. Some unsolved aspects of Nitinol. *Materials Science and Engineering: A*, 438, pp. 69–74.

Duerig, T., 2012. Shape memory alloys. In *ASM Handbook, Volume 23, Materials for Medical Devices* (pp. 237–250). ASM International.

Duernberger, E., MacLeod, C., Lines, D., Loukas, C. and Vasilev, M., 2022. Adaptive optimisation of multi-aperture ultrasonic phased array imaging for increased inspection speeds of wind turbine blade composite panels. *NDT & E International*, 132, p. 102725.

Eaton-Evans, J., Dulieu-Barton, J.M., Little, E.G. and Brown, I.A., 2008. Observations during mechanical testing of Nitinol. *Proceedings of the Institution of Mechanical Engineers, Part C: Journal of Mechanical Engineering Science*, 222(2), pp. 97–105.

EchoNous. 2024. https://echonous.com/

Ensminger, D. and Bond, L.J., 2024. *Ultrasonics: Fundamentals, Technologies, and Applications*. CRC Press.

Favier, D., Liu, Y., Orgeas, L., Sandel, A., Debove, L. and Comte-Gaz, P., 2006. Influence of thermomechanical processing on the superelastic properties of a Ni-rich Nitinol shape memory alloy. *Materials Science and Engineering: A*, 429(1–2), pp. 130–136.

Feeney, A., 2014. Nitinol cymbal transducers for tuneable ultrasonic devices (Doctoral dissertation, University of Glasgow).

Feeney, A. and Lucas, M., 2014. Smart cymbal transducers with nitinol end caps tunable to multiple operating frequencies. *IEEE Transactions on Ultrasonics, Ferroelectrics, and Frequency Control*, 61(10), pp. 1709–1719.

Feeney, A. and Lucas, M., 2016. Differential scanning calorimetry of superelastic Nitinol for tunable cymbal transducers. *Journal of Intelligent Material Systems and Structures*, 27(10), pp. 1376–1387.

Feeney, A. and Lucas, M., 2018. A comparison of two configurations for a dual-resonance cymbal transducer. *IEEE Transactions on Ultrasonics, Ferroelectrics, and Frequency Control*, 65(3), pp. 489–496.

Feeney, A., Kang, L. and Dixon, S., 2017a. Nonlinearity in the dynamic response of flexural ultrasonic transducers. *IEEE Sensors Letters*, 2(1), pp. 1–4.

Feeney, A., Kang, L., Rowlands, G. and Dixon, S., 2017b, December. Dynamic characteristics of flexural ultrasonic transducers. In *Proceedings of Meetings on Acoustics* (Vol. 32, No. 1). AIP Publishing.

Feeney, A., Kang, L., Rowlands, G. and Dixon, S., 2018a. The dynamic performance of flexural ultrasonic transducers. *Sensors*, 18(1), p. 270.

Feeney, A., Kang, L. and Dixon, S., 2018b. High-frequency measurement of ultrasound using flexural ultrasonic transducers. *IEEE Sensors Journal*, 18(13), pp. 5238–5244.

Feeney, A., Kang, L., Rowlands, G., Zhou, L. and Dixon, S., 2019a. Dynamic nonlinearity in piezoelectric flexural ultrasonic transducers. *IEEE Sensors Journal*, 19(15), pp. 6056–6066.

Feeney, A., Kang, L., Somerset, W.E. and Dixon, S., 2019b. The influence of air pressure on the dynamics of flexural ultrasonic transducers. *Sensors*, 19(21), p. 4710.

Feeney, A., Kang, L., Somerset, W.E. and Dixon, S., 2020. Venting in the comparative study of flexural ultrasonic transducers to improve resilience at elevated environmental pressure levels. *IEEE Sensors Journal*, 20(11), pp. 5776–5784.

Feeney, A., Somerset, W.E., Adams, S., Hafezi, M., Kang, L. and Dixon, S., 2023. Measurement stability of oil filled flexural ultrasonic transducers across sequential in-situ pressurization cycles. *IEEE Sensors Journal*, 24, pp. 4281–4289.

Fenton, P., Harrington, F. and Westhaver, P., 2003. Disposable ultrasonic soft tissue cutting and coagulation systems. US Patent 2003/0212332 A1.

Fernandes, F.B., Mahesh, K.K. and dos Santos Paula, A., 2013. Thermomechanical treatments for Ni-Ti alloys. In *Shape Memory Alloys: Processing, Characterization and Applications* (pp. 3–26). https://books.google.co.uk/books/about/Thermomechanical_Treatments_for_Ni_Ti_Al.html?id=xtSqzQEACAAJ&redir_esc=y

Feynman, R., 1963. *Lectures in Physics*, Volume 1, Chapter 47: *Sound. The Wave Equation*. Caltech.

Field, G.S., 1931. Velocity of sound in cylindrical rods. *Canadian Journal of Research*, 5(6), pp. 619–624.

Flatau, A.B., Dapino, M.J. and Calkins, F.T., 2000. High bandwidth tunability in a smart vibration absorber. *Journal of Intelligent Material Systems and Structures*, 11(12), pp. 923–929.

Flores-Méndez, J., Pérez Cuapio, R., Bueno Avendaño, C., Hernández-Ordoñez, M., Aparicio Razo, M., Candia García, F. C., Ambrosio Lázaro, R. and Zenteno-Mateo, B., 2020. The perspective of a homogenization approach for effective local and non-local response of the elastic wave properties of phononic metamaterials. *Advanced Materials Letters*, 11(12), pp. 1–7.

Gallardo Fuentes, J.M., Gümpel, P. and Strittmatter, J., 2002. Phase change behavior of nitinol shape memory alloys. *Advanced Engineering Materials*, 4(7), pp. 437–452.

Gallego-Juárez, J.A., Graff, K.F. and Lucas, M. eds., 2023a. *Power Ultrasonics: Applications of High-Intensity Ultrasound*. Woodhead Publishing.

Gallego-Juárez, J.A., Rodríguez, G., Riera, E., Andrés, R.R. and Cardoni, A., 2023b. Power ultrasonic transducers with vibrating plate radiators. In *Power Ultrasonics* (pp. 109–130). Woodhead Publishing.

Gandomzadeh, D., Rohani, A. and Abbaspour-Fard, M.H., 2023. Future of ultrasonic transducers: How machine learning is driving innovation. *Industrial & Engineering Chemistry Research*, 62(49), pp. 21222–21236.

Gao, N., Starink, M.J. and Langdon, T.G., 2009. Using differential scanning calorimetry as an analytical tool for ultrafine grained metals processed by severe plastic deformation. *Materials Science and Technology*, 25(6), pp. 687–698.

Gao, N., Zhang, Z., Deng, J., Guo, X., Cheng, B. and Hou, H., 2022. Acoustic metamaterials for noise reduction: A review. *Advanced Materials Technologies*, 7(6), p. 2100698.

Gardiner, A., Daly, P., Domingo-Roca, R., Windmill, J.F., Feeney, A. and Jackson-Camargo, J.C., 2021. Additive manufacture of small-scale metamaterial structures for acoustic and ultrasonic applications. *Micromachines*, 12(6), p. 634.

Gil, F.J. and Planell, J.A., 1998. Shape memory alloys for medical applications. *Proceedings of the Institution of Mechanical Engineers, Part H: Journal of Engineering in Medicine*, 212(6), pp. 473–488.

Goldman, R., 1962. *Ultrasonic Technology*. Reinhold Publishing Corporation.

Grabec, T., Sedlák, P., Zoubková, K., Ševčík, M., Janovská, M., Stoklasová, P. and Seiner, H., 2021. Evolution of elastic constants of the NiTi shape memory alloy during a stress-induced martensitic transformation. *Acta Materialia*, 208, p. 116718.

Graff, K.F., 1981. A history of ultrasonics. In *Physical Acoustics* (Vol. 15, pp. 1–97). Academic Press.

Gross, D., Coutier, C., Legros, M., Bouakaz, A. and Certon, D., 2015. A cMUT probe for ultrasound-guided focused ultrasound targeted therapy. *IEEE Transactions on Ultrasonics, Ferroelectrics, and Frequency Control*, 62(6), pp. 1145–1160.

Guo, Y., Klink, A., Fu, C. and Snyder, J., 2013. Machinability and surface integrity of Nitinol shape memory alloy. *CIRP Annals*, 62(1), pp. 83–86.

Hagemann, R., Noelke, C., Rau, T., Kaierle, S., Overmeyer, L., Wesling, V. and Wolkers, W., 2015. Design, processing, and characterization of nickel titanium micro-actuators for medical implants. *Journal of Laser Applications*, 27(S2), p. S29203.

Hall, T., 1994. Joint, a laminate, and a method of preparing a nickel-titanium alloy member surface for bonding to another layer of metal. Patent Number US 5354623.

Harkness, P., McRobb, M., Lützkendorf, P., Milligan, R., Feeney, A. and Clark, C., 2014. Development status of AEOLDOS–A deorbit module for small satellites. *Advances in Space Research*, 54(1), pp. 82–91.

Hasegawa, K. and Shinoda, H., 2018. Aerial vibrotactile display based on multiunit ultrasound phased array. *IEEE Transactions on Haptics*, 11(3), pp. 367–377.

Hernández, J.F., 2016. Synthesis and characterization of biodegradable thermoplastic elastomers for medical applications: Novel copolyesters as an alternative to polylactide and poly (e-caprolactone) (Doctoral dissertation, Universidad del País Vasco-Euskal Herriko Unibertsitatea).

Hirao, M. and Ogi, H., 2013. *EMATs for Science and Industry: Noncontacting Ultrasonic Measurements*. Springer Science & Business Media.

Hirao, M. and Ogi, H., 2017. Introduction: Noncontact ultrasonic measurements. In *Electromagnetic Acoustic Transducers: Noncontacting Ultrasonic Measurements Using EMATs* (pp. 1–11).

Hodgson, D. and Russell, S., 2000. Nitinol melting, manufacture and fabrication. *Minimally Invasive Therapy & Allied Technologies*, 9(2), pp. 61–65.

Howell, L., Ingram, N., Lapham, R., Morrell, A. and McLaughlan, J.R., 2024. Deep learning for real-time multi-class segmentation of artefacts in lung ultrasound. *Ultrasonics*, 140, p. 107251.

Hu, Y., Guo, Z., Ragonese, A., Zhu, T., Khuje, S., Li, C., Grossman, J.C., Zhou, C., Nouh, M. and Ren, S., 2020. A 3D-printed molecular ferroelectric metamaterial. *Proceedings of the National Academy of Sciences*, 117(44), pp. 27204–27210.

Huang, X. and Liu, Y., 2001. Effect of annealing on the transformation behavior and superelasticity of NiTi shape memory alloy. *Scripta Materialia*, 45(2), pp. 153–160.

Huang, H.H. and Sun, C.T., 2011. Theoretical investigation of the behavior of an acoustic metamaterial with extreme Young's modulus. *Journal of the Mechanics and Physics of Solids*, 59(10), pp. 2070–2081.

Hyun, D. and Dahl, J.J., 2020. Effects of motion on correlations of pulse-echo ultrasound signals: Applications in delay estimation and aperture coherence. *The Journal of the Acoustical Society of America*, 147(3), pp. 1323–1332.

Inui, A., Mifune, Y., Nishimoto, H., Mukohara, S., Fukuda, S., Kato, T., Furukawa, T., Tanaka, S., Kusunose, M., Takigami, S. and Ehara, Y., 2023. Detection of elbow OCD in the ultrasound image by artificial intelligence using YOLOv8. *Applied Sciences*, 13(13), p. 7623.

Jain, A.K., Sharma, A.K., Khandekar, S. and Bhattacharya, B., 2020. Shape memory alloy-based sensor for two-phase flow detection. *IEEE Sensors Journal*, 20(23), pp. 14209–14217.

Jalaeefar, A. and Asgarian, B., 2013. Experimental investigation of mechanical properties of nitinol, structural steel, and their hybrid component. *Journal of Materials in Civil Engineering*, 25(10), pp. 1498–1505.

Jan, M., 2023. About the physiology of hearing. *Journal of Acute Care & Trauma Surgery*, 1(1), pp. 1–4.

Jani, J.M., Leary, M., Subic, A. and Gibson, M.A., 2014. A review of shape memory alloy research, applications and opportunities. *Materials & Design (1980–2015)*, 56, pp. 1078–1113.

Ji, G. and Huber, J., 2022. Recent progress in acoustic metamaterials and active piezoelectric acoustic metamaterials-a review. *Applied Materials Today*, 26, p. 101260.

Jiang, C., Li, Z., Zhang, Z. and Wang, S., 2023. A new design to Rayleigh wave EMAT based on spatial pulse compression. *Sensors*, 23(8), p. 3943.

Joshi, S.V., Sadeghpour, S. and Kraft, M., 2023. Polyimide-on-silicon 2D piezoelectric micromachined ultrasound transducer (PMUT) array. *Sensors*, 23(10), p. 4826.

Jung, J., Lee, W., Kang, W., Shin, E., Ryu, J. and Choi, H., 2017. Review of piezoelectric micromachined ultrasonic transducers and their applications. *Journal of Micromechanics and Microengineering*, 27(11), p. 113001.

Kang, L., Feeney, A. and Dixon, S., 2017, December. Flow measurement based on two-dimensional flexural ultrasonic phased arrays. In *Proceedings of Meetings on Acoustics* (Vol. 32, No. 1). AIP Publishing.

Kang, L., Feeney, A., Su, R., Lines, D., Ramadas, S.N., Rowlands, G. and Dixon, S., 2019. Flow velocity measurement using a spatial averaging method with two-dimensional flexural ultrasonic array technology. *Sensors*, 19(21), p. 4786.

Kang, L., Feeney, A. and Dixon, S., 2020. The high frequency flexural ultrasonic transducer for transmitting and receiving ultrasound in air. *IEEE Sensors Journal*, 20(14), pp. 7653–7660.

Kastner, O., 2012. *First Principles Modelling of Shape Memory Alloys: Molecular Dynamics Simulations* (Vol. 163). Springer Science & Business Media.

Kennedy, A., Brame, J., Rycroft, T., Wood, M., Zemba, V., Weiss Jr, C., Hull, M., Hill, C., Geraci, C. and Linkov, I., 2019. A definition and categorization system for advanced materials: The foundation for risk-informed environmental health and safety testing. *Risk Analysis*, 39(8), pp. 1783–1795.

Ketsuwan, P., Prasatkhetragarn, A., Triamnuk, N., Huang, C.C., Ngamjarurojana, A., Ananta, S., Cann, D.P. and Yimnirun, R., 2009. Electrical conductivity and dielectric and ferroelectric properties of chromium doped lead zirconate titanate ceramic. *Ferroelectrics*, 382(1), pp. 49–55.

Khaing, Z.Z., Cates, L.N., DeWees, D.M., Hannah, A., Mourad, P., Bruce, M. and Hofstetter, C.P., 2018. Contrast-enhanced ultrasound to visualize hemodynamic changes after rodent spinal cord injury. *Journal of Neurosurgery: Spine*, 29(3), pp. 306–313.

Khan, U., Khan, N.Z. and Gulati, J., 2017. Ultrasonic welding of bi-metals: Optimizing process parameters for maximum tensile-shear strength and plasticity of welds. *Procedia Engineering*, 173, pp. 1447–1454.

Khan, A., Mineo, C., Dobie, G., Macleod, C.N. and Pierce, S.G., 2019. Introducing adaptive vision-guided robotic non-destructive inspection. In *Review of Progress in Quantitative Nondestructive Evaluation*. https://scholar.google.co.uk/scholar?q= Introducing+adaptive+vision-guided+robotic+non-destructive+inspection&hl=en& as_sdt=0&as_vis=1&oi=scholart

Khuri-Yakub, B.T. and Oralkan, Ö., 2011. Capacitive micromachined ultrasonic transducers for medical imaging and therapy. *Journal of Micromechanics and Microengineering*, 21(5), p. 054004.

Kim, H.C., Yoo, Y.I. and Lee, J.J., 2008. Development of a NiTi actuator using a two-way shape memory effect induced by compressive loading cycles. *Sensors and Actuators A: Physical*, 148(2), pp. 437–442.

Kim, J.W., Hwang, G., Lee, S.J., Kim, S.H. and Wang, S., 2022. Three-dimensional acoustic metamaterial Luneburg lenses for broadband and wide-angle underwater ultrasound imaging. *Mechanical Systems and Signal Processing*, 179, p. 109374.

Knorr, D., Zenker, M., Heinz, V. and Lee, D.U., 2004. Applications and potential of ultrasonics in food processing. *Trends in Food Science & Technology*, 15(5), pp. 261–266.

Korolev, I., Aliev, T.A., Orlova, T., Ulasevich, S.A., Nosonovsky, M. and Skorb, E.V., 2022. When bubbles are not spherical: Artificial intelligence analysis of ultrasonic cavitation bubbles in solutions of varying concentrations. *The Journal of Physical Chemistry B*, 126(16), pp. 3161–3169.

Kuhn, G. and Jordan, L., 2002. Fatigue and mechanical properties of nickel-titanium endodontic instruments. *Journal of Endodontics*, 28(10), pp. 716–720.

Kumar, P. and Huang, S., 2023. Effect of pre-strain, temperature, and time on the recovery behavior of Nitinol. *Shape Memory and Superelasticity*, 9(1), pp. 5–10.

Kumar, P.K. and Lagoudas, D.C., 2008. Introduction to shape memory alloys. In *Shape Memory Alloys: Modeling and Engineering Applications* (pp. 1–51). Springer US.

Leissa, A.W., 1969. *Vibration of Plates* (Vol. 160). Scientific and Technical Information Division, National Aeronautics and Space Administration.

Lesota, A., Sibirev, A., Rubanik, V., Rubanik Jr, V., Resnina, N. and Belyaev, S., 2019. Initiation of the shape memory effect by temperature variation or ultrasonic vibrations in the NiTi shape memory alloy after different preliminary deformation. *Sensors and Actuators A: Physical*, 286, pp. 1–3.

Li, D.Z. and Feng, Z.C., 1997, June. Dynamic properties of pseudoelastic shape memory alloys. In *Smart Structures and Materials 1997: Smart Structures and Integrated Systems* (Vol. 3041, pp. 715–725). SPIE.

Li, M., Wen, Y., Li, P. and Yang, J., 2011, October. A magnetostrictive/piezoelectric laminate transducer based vibration energy harvester with resonance frequency tunability. In *SENSORS, 2011 IEEE* (pp. 1768–1771). IEEE.

Li, J., Ren, W., Fan, G. and Wang, C., 2017. Design and fabrication of piezoelectric micromachined ultrasound transducer (pMUT) with partially-etched ZnO film. *Sensors*, 17(6), p. 1381.

Li, M. and Hayward, G., 2019, May. Adaptive array processing for enhanced ultrasonic non-destructive evaluation (NDE) and imaging. In *AIP Conference Proceedings* (Vol. 2102, No. 1). AIP Publishing.

Li, Y., Yu, B., Wang, B., Lee, T.H. and Banu, M., 2020. Online quality inspection of ultrasonic composite welding by combining artificial intelligence technologies with welding process signatures. *Materials & Design*, 194, p. 108912.

Li, W., Wang, J., Qin, X., Jing, G. and Liu, X., 2024. Artificial intelligence-aided detection of rail defects based on ultrasonic imaging data. *Proceedings of the Institution of Mechanical Engineers, Part F: Journal of Rail and Rapid Transit*, 238(1), pp. 118–127.

Li, X., Stritch, T. and Lucas, M., 2019, October. Design of miniature ultrasonic surgical devices. In *2019 IEEE International Ultrasonics Symposium (IUS)* (pp. 2641–2644). IEEE.

Li, X., Stritch, T., Manley, K. and Lucas, M., 2021. Limits and opportunities for miniaturizing ultrasonic surgical devices based on a Langevin transducer. *IEEE Transactions on Ultrasonics, Ferroelectrics, and Frequency Control*, 68(7), pp. 2543–2553.

Lindberg, J.F., 1983. Parametric dual mode transducer. US Patent 4,373,143.

Ling, H.C. and Kaplow, R., 1981. Stress-induced shape changes and shape memory in the R and martensite transformations in equiatomic NiTi. *Metallurgical Transactions A*, 12, pp. 2101–2111.

Liu, X., Wang, Y., Yang, D. and Qi, M., 2008. The effect of ageing treatment on shape-setting and superelasticity of a nitinol stent. *Materials Characterization*, 59(4), pp. 402–406.

Liu, Y. and Xiang, H., 1998. Apparent modulus of elasticity of near-equiatomic NiTi. *Journal of Alloys and Compounds*, 270(1–2), pp. 154–159.

Liu, J., Guo, H. and Wang, T., 2020. A review of acoustic metamaterials and phononic crystals. *Crystals*, 10(4), p. 305.

Liu, Y., Hafezi, M. and Feeney, A., 2023a. Fabrication and dynamic characterisation of a Nitinol Langevin transducer. In *17th International Conference on Advances in Experimental Mechanics*, Glasgow, UK, 30 August–01 September 2023.

Liu, Y., Hafezi, M. and Feeney, A., 2023b, September. Active modal coupling of a Nitinol Langevin transducer. In *2023 IEEE International Ultrasonics Symposium (IUS)* (pp. 1–4). IEEE.

Liu, Y., Nguyen, L.T.K., Li, X. and Feeney, A., 2024a. A Timoshenko-Ehrenfest beam model for simulating Langevin transducer dynamics. *Applied Mathematical Modelling*, 131, pp. 363–380.

Liu, Y., Hafezi, M. and Feeney, A., 2024b. A cascaded Nitinol Langevin transducer for resonance stability at elevated temperatures. *Ultrasonics*, 137, p. 107201.

Lorraine, P.W., 1995. Ultrasonic transducer with selectable center frequency. US Patent 5,381,068.

Lucas, M., Mathieson, A. and Cleary, R., 2023. Ultrasonic cutting for surgical applications. In *Power Ultrasonics* (pp. 617–635). Woodhead Publishing.

Luijten, B., Cohen, R., De Bruijn, F.J., Schmeitz, H.A., Mischi, M., Eldar, Y.C. and Van Sloun, R.J., 2020. Adaptive ultrasound beamforming using deep learning. *IEEE Transactions on Medical Imaging*, 39(12), pp. 3967–3978.

Ma, G. and Sheng, P., 2016. Acoustic metamaterials: From local resonances to broad horizons. *Science Advances*, 2(2), p. e1501595.

Macleod, C.N., Dobie, G., Pierce, S.G., Summan, R. and Morozov, M., 2016. Machining-based coverage path planning for automated structural inspection. *IEEE Transactions on Automation Science and Engineering*, 15(1), pp. 202–213.

Madison, T.C. and Frey, H.G., 1976. Multiple-frequency transducer. US Patent 3,952,216.

Malik, P., Pathania, M. and Rathaur, V.K., 2019. Overview of artificial intelligence in medicine. *Journal of Family Medicine and Primary Care*, 8(7), p. 2328.

Malik, V., Srivastava, S., Gupta, S., Sharma, V., Vishnoi, M. and Mamatha, T.G., 2021. A novel review on shape memory alloy and their applications in extraterrestrial roving missions. *Materials Today: Proceedings*, 44, pp. 4961–4965.

Manwar, R. and Chowdhury, S., 2016. Experimental analysis of bisbenzocyclobutene bonded capacitive micromachined ultrasonic transducers. *Sensors*, 16(7), p. 959.

Martinho, L.M., Kubrusly, A.C., Kang, L. and Dixon, S., 2022. Enhancement of the unidirectional radiation pattern of shear horizontal ultrasonic waves generated by side-shifted periodic permanent magnets electromagnetic acoustic transducers with multiple rows of magnets. *IEEE Sensors Journal*, 22(8), pp. 7637–7644.

Martins, R.M.S., Schell, N., Silva, R.J.C. and Fernandes, F.B., 2005. Structural in situ studies of shape memory alloy (SMA) Ni–Ti thin films. *Nuclear Instruments and Methods in Physics Research Section B: Beam Interactions with Materials and Atoms*, 238(1–4), pp. 319–322.

Maruyama, T., Hirata, H., Furukawa, T. and Maruo, S., 2020. Multi-material microstereolithography using a palette with multicolor photocurable resins. *Optical Materials Express*, 10(10), pp. 2522–2532.

Mason, W.P., 1948. *Electromechanical Transducers and Wave Filters*. Van Nostrand.

Mason, W.P., 1976, September. Sonics and ultrasonics: Early history and applications. In *1976 Ultrasonics Symposium* (pp. 610–617). IEEE.

Mathieson, A., Cardoni, A., Cerisola, N. and Lucas, M., 2013. The influence of piezoceramic stack location on nonlinear behavior of Langevin transducers. *IEEE Transactions on Ultrasonics, Ferroelectrics, and Frequency Control*, 60(6), pp. 1126–1133.

Mathieson, A., Feeney, A., Tweedie, A. and Lucas, M., 2015a, October. Ultrasonic biopsy needle based on the class IV flextensional configuration. In *2015 IEEE International Ultrasonics Symposium (IUS)* (pp. 1–4). IEEE.

Mathieson, A., Cardoni, A., Cerisola, N. and Lucas, M., 2015b. Understanding nonlinear vibration behaviours in high-power ultrasonic surgical devices. *Proceedings of the Royal Society A: Mathematical, Physical and Engineering Sciences*, 471(2176), p. 20140906.

Mathieson, A. and DeAngelis, D.A., 2015. Analysis of lead-free piezoceramic-based power ultrasonic transducers for wire bonding. *IEEE Transactions on Ultrasonics, Ferroelectrics, and Frequency Control*, 63(1), pp. 156–164.

Mathieson, A., Tyson, B. and Bond, A., 2019. The demonstration of additive manufacture in power ultrasonic and sonar transducers. *IEEE Transactions on Ultrasonics, Ferroelectrics, and Frequency Control*, 67(4), pp. 817–824.

Mauchamp, P. and Flesch, A., 2003. Multi-purpose ultrasonic slotted array transducer. US Patent 6,537,224 B2.

McKelvey, A.L. and Ritchie, R.O., 2000. On the temperature dependence of the superelastic strength and the prediction of the theoretical uniaxial transformation strain in Nitinol. *Philosophical Magazine A*, 80(8), pp. 1759–1768.

Martinussen, H., Aksnes, A., Leirset, E. and Engan, H.E., 2009. CMUT characterization by interferometric and electric measurements. *IEEE Transactions on Ultrasonics, Ferroelectrics, and Frequency Control*, 56(12), pp. 2711–2721.

Matsukawa, M., Choi, P.K., Nakamura, K., Ogi, H. and Hasegawa, H., 2022. *Ultrasonics: Physics and Applications*. IOP Publishing.

McNaney, J.M., Imbeni, V., Jung, Y., Papadopoulos, P. and Ritchie, R.O., 2003. An experimental study of the superelastic effect in a shape-memory Nitinol alloy under biaxial loading. *Mechanics of Materials*, 35(10), pp. 969–986.

Mertmann, M. and Vergani, G., 2008. Design and application of shape memory actuators. *The European Physical Journal Special Topics*, 158(1), pp. 221–230.

Meyer Jr, R.J. and Newnham, R.E., 2000. Flextensional transducers with shape memory caps for tunable devices. *Journal of Intelligent Material Systems and Structures*, 11(3), pp. 199–205.

Miyazaki, S. and Otsuka, K., 1986. Deformation and transition behavior associated with the R-phase in Ti-Ni alloys. *Metallurgical Transactions A*, 17, pp. 53–63.

Morisaki, T., Fujiwara, M., Makino, Y. and Shinoda, H., 2023. Noncontact haptic rendering of static contact with convex surface using circular movement of ultrasound focus on a finger pad. In *IEEE Transactions on Haptics* (pp. 1–12). IEEE.

Mozaffarzadeh, M., Minonzio, C., De Jong, N., Verweij, M.D., Hemm, S. and Daeichin, V., 2020. Lamb waves and adaptive beamforming for aberration correction in medical ultrasound imaging. *IEEE Transactions on Ultrasonics, Ferroelectrics, and Frequency Control*, 68(1), pp. 84–91.

Mozumi, M. and Hasegawa, H., 2018. Adaptive beamformer combined with phase coherence weighting applied to ultrafast ultrasound. *Applied Sciences*, 8(2), p. 204.

Narayanan, M., Schwartz, R.W. and Zhou, D., 2007. Stress-biased cymbals using shape memory alloys. *Journal of the American Ceramic Society*, 90(4), pp. 1122–1129.

Neuweiler, G., Bruns, V. and Schuller, G., 1980. Ears adapted for the detection of motion, or how echolocating bats have exploited the capacities of the mammalian auditory system. *The Journal of the Acoustical Society of America*, 68(3), pp. 741–753.

Newnham, R.E., 1992, May. Piezoelectric sensors and actuators: Smart materials. In *Proceedings of the 1992 IEEE Frequency Control Symposium* (pp. 513–524). IEEE.

Newnham, R.E., Xu, Q.C. and Blaszkiewicz, M., 1992. Frequency agile sonic transducer. US Patent 5,166,907.

Ning, G., Zhang, X., Zhang, Q., Wang, Z. and Liao, H., 2020. Real-time and multimodality image-guided intelligent HIFU therapy for uterine fibroid. *Theranostics*, 10(10), p. 4676.

Norouzi, A., Hamedi, M. and Adineh, V.R., 2012. Strength modeling and optimizing ultrasonic welded parts of ABS-PMMA using artificial intelligence methods. *The International Journal of Advanced Manufacturing Technology*, 61, pp. 135–147.

Watson, E.S. and O'Neill, M.J., 1962. Differential microcalorimeter. US Patent 3,263,484.

Olbricht, J., Yawny, A., Pelegrina, J.L., Dlouhy, A. and Eggeler, G., 2011. On the stress-induced formation of R-phase in ultra-fine-grained Ni-rich NiTi shape memory alloys. *Metallurgical and Materials Transactions A*, 42, pp. 2556–2574.

Paul, S., Mandal, S. and Singh, M.S., 2021. Noise adaptive beamforming for linear array photoacoustic imaging. *IEEE Transactions on Instrumentation and Measurement*, 70, pp. 1–11.

Paula, A.S., Canejo, J.P.H.G., Martins, R.M.S. and Fernandes, F.B., 2004. Effect of thermal cycling on the transformation temperature ranges of a Ni–Ti shape memory alloy. *Materials Science and Engineering: A*, 378(1–2), pp. 92–96.

Pelenis, D., Barauskas, D., Vanagas, G., Dzikaras, M. and Viržonis, D., 2019. CMUT-based biosensor with convolutional neural network signal processing. *Ultrasonics*, 99, p. 105956.

Pelton, A.R., DiCello, J. and Miyazaki, S., 2000. Optimisation of processing and properties of medical grade nitinol wire. *Minimally Invasive Therapy & Allied Technologies*, 9(1), pp. 107–118.

Pelton, A.R., Russell, S.M. and DiCello, J., 2003. The physical metallurgy of nitinol for medical applications. *JOM*, 55(5), pp. 33–37.

Perry, K.E. and Labossiere, P.E., 2004. Phase transformations in nitinol and challenges for numerical modeling. *Medical Device Materials II*, pp. 131–134.

Petcher, P.A., Potter, M.D.G. and Dixon, S., 2014. A new electromagnetic acoustic transducer (EMAT) design for operation on rail. *NDT & E International*, 65, pp. 1–7.

Peters, C., Maurath, D., Schock, W. and Manoli, Y., 2008. Novel electrically tunable mechanical resonator for energy harvesting. In *Proceedings of PowerMEMS* (pp. 253–256). https://www.powermems.org/about/previous_conferences.html

Petrover, K. and Baz, A., 2020. Finite element modeling of one-dimensional nonreciprocal acoustic metamaterial with anti-parallel diodes. *The Journal of the Acoustical Society of America*, 148(1), pp. 334–346.

Poncet, P.P., 2000. Nitinol medical device design considerations. *Strain*, 2(4), p. 6.

Porzio, R., 2009. Multiple frequency sonar transducer. US Patent 7,535,801 B1.

Price, A. and Long, B., 2018, October. Fibonacci spiral arranged ultrasound phased array for mid-air haptics. In *2018 IEEE International Ultrasonics Symposium (IUS)* (pp. 1–4). IEEE.

Pyo, S., Afzal, M.S., Lim, Y., Lee, S. and Roh, Y., 2021. Design of a wideband tonpilz transducer comprising non-uniform piezoceramic stacks with equivalent circuits. *Sensors*, 21(8), p. 2680.

OED Online. 2024. https://www.oed.com/

Oxford Reference. 2024. https://www.oxfordreference.com/

Ranjan, B.S.C., Vikranth, H.N. and Ghosal, A., 2013. A novel prevailing torque threaded fastener and its analysis. *Journal of Mechanical Design*, 135(10), p. 101007.

Rajagopalan, S., Little, A.L., Bourke, M.A.M. and Vaidyanathan, R., 2005. Elastic modulus of shape-memory NiTi from in situ neutron diffraction during macroscopic loading, instrumented indentation, and extensometry. *Applied Physics Letters*, 86(8), p. 081901.

Richardson, E.G., 1962. *Ultrasonic Physics*. Elsevier.

Robertson, S.W. and Ritchie, R.O., 2008. A fracture-mechanics-based approach to fracture control in biomedical devices manufactured from superelastic Nitinol tube. *Journal of Biomedical Materials Research Part B: Applied Biomaterials*, 84(1), pp. 26–33.

Rojas, S.S., Tridandapani, S. and Lindsey, B.D., 2021. A thin transducer with integrated acoustic metamaterial for cardiac CT imaging and gating. *IEEE Transactions on Ultrasonics, Ferroelectrics, and Frequency Control*, 69(3), pp. 1064–1076.

Rolt, K.D., 1990. History of the flextensional electroacoustic transducer. *The Journal of the Acoustical Society of America*, 87(3), pp. 1340–1349.

Rossing, T.D., 2014. Introduction to acoustics. In *Springer Handbook of Acoustics*. Springer: New York, NY.

Rouffaud, R., Granger, C., Hladky-Hennion, A.C., Thi, M.P. and Levassort, F., 2015. Tonpilz underwater acoustic transducer integrating lead-free piezoelectric material. *Physics Procedia*, 70, pp. 997–1001.

Roy, K., Lee, J.E.Y. and Lee, C., 2023. Thin-film PMUTs: A review of over 40 years of research. *Microsystems & Nanoengineering*, 9(1), p. 95.

RoyChoudhury, S., Rawat, V., Jalal, A.H., Kale, S.N. and Bhansali, S., 2016. Recent advances in metamaterial split-ring-resonator circuits as biosensors and therapeutic agents. *Biosensors and Bioelectronics*, 86, pp. 595–608.

Rubanik Jr, V.V., Rubanik, V.V. and Klubovich, V.V., 2008. The influence of ultrasound on shape memory behavior. *Materials Science and Engineering: A*, 481, pp. 620–622.

Saadat, S., Salichs, J., Noori, M., Hou, Z., Davoodi, H., Bar-On, I., Suzuki, Y. and Masuda, A., 2002. An overview of vibration and seismic applications of NiTi shape memory alloy. *Smart Materials and Structures*, 11(2), p. 218.

Sachdeva, R.C.L., Miyazaki, S. and Dughaish, Z.H., 2001. Nitinol as a biomedical material. In *Encyclopedia of Materials: Science and Technology* (pp. 6155–6160). https://www.sciencedirect.com/science/article/pii/B0080431526010937?via%3Dihub

Sambath, S., Nagaraj, P. and Selvakumar, N., 2011. Automatic defect classification in ultrasonic NDT using artificial intelligence. *Journal of Nondestructive Evaluation*, 30, pp. 20–28.

Sasso, M. and Cohen-Bacrie, C., 2005, March. Medical ultrasound imaging using the fully adaptive beamformer. In *Proceedings. (ICASSP'05). IEEE International Conference on Acoustics, Speech, and Signal Processing, 2005* (Vol. 2, pp. ii–489). IEEE.

Sautto, M., Savoia, A.S., Quaglia, F., Caliano, G. and Mazzanti, A., 2017. A comparative analysis of CMUT receiving architectures for the design optimization of integrated transceiver front ends. *IEEE Transactions on Ultrasonics, Ferroelectrics, and Frequency Control*, 64(5), pp. 826–838.

Schafer, M.E. and Cleary, R., 2023. Ultrasonic surgical devices and procedures. In *Power Ultrasonics* (pp. 539–556). Woodhead Publishing.

Schlun, M., Zipse, A., Dreher, G. and Rebelo, N., 2011. Effects of cyclic loading on the uniaxial behavior of Nitinol. *Journal of Materials Engineering and Performance*, 20, pp. 684–687.

Selvan, K.T. and Janaswamy, R., 2017. Fraunhofer and Fresnel distances: Unified derivation for aperture antennas. *IEEE Antennas and Propagation Magazine*, 59(4), pp. 12–15.

Shabalovskaya, S.A., Tian, H., Anderegg, J.W., Schryvers, D.U., Carroll, W.U. and Van Humbeeck, J., 2009. The influence of surface oxides on the distribution and release of nickel from Nitinol wires. *Biomaterials*, 30(4), pp. 468–477.

Shaw, J.A., 2008. Tips and tricks for characterizing shape memory alloy wire: Part 1— Differential scanning calorimetry and basic phenomena. *Experimental Techniques*, 32(5), pp. 55–62.

Shen, Y.T., Chen, L., Yue, W.W. and Xu, H.X., 2021. Artificial intelligence in ultrasound. *European Journal of Radiology*, 139, p. 109717.

Sherman, C.H. and Butler, J.L., 2007. *Transducers and Arrays for Underwater Sound* (Vol. 4). Springer.

Shim, H. and Roh, Y., 2019. Design and fabrication of a wideband cymbal transducer for underwater sensor networks. *Sensors*, 19(21), p. 4659.

Shimoga, G., Kim, T.H. and Kim, S.Y., 2021. An intermetallic NiTi-based shape memory coil spring for actuator technologies. *Metals*, 11(8), p. 1212.

Shin, D.D., Mohanchandra, K.P. and Carman, G.P., 2004. High frequency actuation of thin film NiTi. *Sensors and Actuators A: Physical*, 111(2–3), pp. 166–171.

Singh, P.K., Kumar, S.D., Patel, D. and Prasad, S.B., 2017. Optimization of vibratory welding process parameters using response surface methodology. *Journal of Mechanical Science and Technology*, 31, pp. 2487–2495.

Smith, S., Li, X., Hafezi, M., Barron, P., Lucas, M. and Feeney, A., 2022, October. Enhanced resolution phase transformations in a Nitinol cymbal ultrasonic device. In *2022 IEEE International Ultrasonics Symposium (IUS)* (pp. 1–4). IEEE.

Somerset, W.E., Feeney, A., Kang, L., Li, Z. and Dixon, S., 2021, September. Oil filled flexural ultrasonic transducers for resilience in environments of elevated pressure. In *2021 IEEE International Ultrasonics Symposium (IUS)* (pp. 1–4). IEEE.

Somerset, W.E., Feeney, A., Kang, L., Li, Z. and Dixon, S., 2022. Design and dynamics of oil filled flexural ultrasonic transducers for elevated pressures. *IEEE Sensors Journal*, 22(13), pp. 12673–12680.

Song, Y., Chen, X., Dabade, V., Shield, T.W. and James, R.D., 2013. Enhanced reversibility and unusual microstructure of a phase-transforming material. *Nature*, 502(7469), pp. 85–88.

Song, Y. and Shen, Y., 2020, November. Steerable unidirectional wave emission from a single piezoelectric transducer using a shape memory alloy composite metasurface. In *ASME International Mechanical Engineering Congress and Exposition* (Vol. 84478, p. V001T01A003). American Society of Mechanical Engineers.

Stansfield, D. and Elliott, A., 2017. *Underwater Electroacoustic Transducers*. Peninsula Publishing.

Stearns, C.M., Erickson, D.J. and Izzo, L.M., 1996. Dual frequency sonar transducer assembly. US Patent 5,515,342.

Stöckel, D., 1995. The shape memory effect-phenomenon, alloys and applications. *Proceedings: Shape Memory Alloys for Power Systems EPRI*, 1, pp. 1–13.

Stöckel, D., 1998. Nitinol-A material with unusual properties. *Endovascular Update*, 1(1), pp. 1–8.

Stoeckel, D., 2000. Nitinol medical devices and implants. *Minimally Invasive Therapy & Allied Technologies*, 9(2), pp. 81–88.

Stoeckel, D., Pelton, A. and Duerig, T., 2004. Self-expanding nitinol stents: Material and design considerations. *European Radiology*, 14(2), pp. 292–301.

Sun, H., Ramuhalli, P. and Jacob, R.E., 2023. Machine learning for ultrasonic nondestructive examination of welding defects: A systematic review. *Ultrasonics*, 127, p. 106854.

Suñol, F., Ochoa, D.A., Suñé, L.R. and García, J.E., 2019. Design and characterization of immersion ultrasonic transducers for pulsed regime applications. *Instrumentation Science & Technology*, 47(2), pp. 213–232.

Synnevag, J.F., Austeng, A. and Holm, S., 2007. Adaptive beamforming applied to medical ultrasound imaging. *IEEE Transactions on Ultrasonics, Ferroelectrics, and Frequency Control*, 54(8), pp. 1606–1613.

Szabo, T.L., 2004. *Diagnostic Ultrasound Imaging: Inside Out*. Academic Press.

Tadaki, T., Nakata, Y. and Shimizu, K.I., 1987. Thermal cycling effects in an aged Ni-rich Ti–Ni shape memory alloy. *Transactions of the Japan Institute of Metals*, 28(11), pp. 883–890.

Thomasová, M., Sedlák, P., Seiner, H., Janovská, M., Kabla, M., Shilo, D. and Landa, M., 2015. Young's moduli of sputter-deposited NiTi films determined by resonant ultrasound spectroscopy: Austenite, R-phase, and martensite. *Scripta Materialia*, 101, pp. 24–27.

Thompson, S.C., 1986a. Broadband radial vibrator transducer with multiple resonant frequencies. US Patent 4,604,542.

Thompson, S.C., 1986b. Broadband multi-resonant longitudinal vibrator transducer. US Patent 4,633,119.

Thompson, S.A., 2000. An overview of nickel–titanium alloys used in dentistry. *International Endodontic Journal*, 33(4), pp. 297–310.

Tressler, J.F. and Newnham, R.E., 1997. Doubly resonant cymbal-type transducers. *IEEE Transactions on Ultrasonics, Ferroelectrics, and Frequency Control*, 44(5), pp. 1175–1177.

Triki, H. and Kruglov, V.I., 2022. Chirped periodic and solitary waves in nonlinear negative index materials. *Optics Communications*, 502, p. 127409.

Uchil, J., Mohanchandra, K.P., Mahesh, K.K. and Kumara, K.G., 1998. Thermal and electrical characterization of R-phase dependence on heat-treat temperature in Nitinol. *Physica B: Condensed Matter*, 253(1–2), pp. 83–89.

Uhlig, S., Alkhasli, I., Schubert, F., Tschöpe, C. and Wolff, M., 2023. A review of synthetic and augmented training data for machine learning in ultrasonic non-destructive evaluation. *Ultrasonics*, 134, p. 107041.

Van Humbeeck, J., 1999. Non-medical applications of shape memory alloys. *Materials Science and Engineering: A*, 273, pp. 134–148.

Van Sloun, R.J., Cohen, R. and Eldar, Y.C., 2019. Deep learning in ultrasound imaging. *Proceedings of the IEEE*, 108(1), pp. 11–29.

Vasilev, M., MacLeod, C.N., Loukas, C., Javadi, Y., Vithanage, R.K., Lines, D., Mohseni, E., Pierce, S.G. and Gachagan, A., 2021. Sensor-enabled multi-robot system for automated welding and in-process ultrasonic NDE. *Sensors*, 21(15), p. 5077.

Vidal, N., Asua, E., Feuchtwanger, J., García-Arribas, A., Gutierrez, J. and Barandiaran, J.M., 2008. FEM simulation of the Nitinol wire. *The European Physical Journal Special Topics*, 158, pp. 39–44.

Vilar, Z.T., Grassi, E.N.D., de Oliveira, H.M.R. and de Araujo, C.J., 2023. Use of dynamic mechanical analysers to characterize shape memory alloys: Cautions and considerations from an experimental analysis. *International Journal of Thermophysics*, 44(9), p. 142.

Walker, J., Andani, M.T., Haberland, C. and Elahinia, M., 2014, November. Additive manufacturing of Nitinol shape memory alloys to overcome challenges in conventional Nitinol fabrication. In *ASME International Mechanical Engineering Congress and Exposition* (Vol. 46438, p. V02AT02A037). American Society of Mechanical Engineers.

Wang, D., Wang, X., Chen, S., Li, J., Liang, L. and Liu, Y., 2024. Joint learning of sparse and limited-view guided waves signals for feature reconstruction and imaging. *Ultrasonics*, 137, p. 107200.

Wang, J., Allein, F., Boechler, N., Friend, J. and Vazquez-Mena, O., 2021. Design and fabrication of negative-refractive-index metamaterial unit cells for near-megahertz enhanced acoustic transmission in biomedical ultrasound applications. *Physical Review Applied*, 15(2), p. 024025.

Wang, W., He, Q., Zhang, Z. and Feng, Z., 2022. Adaptive beamforming based on minimum variance (ABF-MV) using deep neural network for ultrafast ultrasound imaging. *Ultrasonics*, 126, p. 106823.

Wang, X.M. and Yue, Z.F., 2007. FEM prediction of the pseudoelastic behavior of NiTi SMA at different temperatures with one temperature testing results. *Computational Materials Science*, 39(3), pp. 697–704.

Wegel, R.L. and Walther, H., 1935. Internal dissipation in solids for small cyclic strains. *Physics*, 6(4), pp. 141–157.

Williams, K.A., Chiu, G.T. and Bernhard, R.J., 1999, June. Passive-adaptive vibration absorbers using shape memory alloys. In *Smart Structures and Materials 1999: Smart Structures and Integrated Systems* (Vol. 3668, pp. 630–641). SPIE.

Wong, G.S., 1986. Speed of sound in standard air. *The Journal of the Acoustical Society of America*, 79(5), pp. 1359–1366.

Wu, Y.C. and Feng, J.W., 2018. Development and application of artificial neural network. *Wireless Personal Communications*, 102, pp. 1645–1656.

Xie, M., Huan, Q. and Li, F., 2020. Quick and repeatable shear modulus measurement based on torsional resonance using a piezoelectric torsional transducer. *Ultrasonics*, 103, p. 106101.

Xu, L. and Wang, R., 2010. The effect of annealing and cold-drawing on the super-elasticity of the Ni-Ti shape memory alloy wire. *Modern Applied Science*, 4(12), p. 109.

Yüksel, N., Börklü, H.R., Sezer, H.K. and Canyurt, O.E., 2023. Review of artificial intelligence applications in engineering design perspective. *Engineering Applications of Artificial Intelligence*, 118, p. 105697.

Zangeneh-Nejad, F. and Fleury, R., 2019. Active times for acoustic metamaterials. *Reviews in Physics*, 4, p. 100031.

Zarnetta, R., Takahashi, R., Young, M.L., Savan, A., Furuya, Y., Thienhaus, S., Maaß, B., Rahim, M., Frenzel, J., Brunken, H. and Chu, Y.S., 2010. Identification of quaternary shape memory alloys with near-zero thermal hysteresis and unprecedented functional stability. *Advanced Functional Materials*, 20(12), pp. 1917–1923.

Zeng, H.C., Luo, C.R., Chen, H.J., Zhai, S.L., Ding, C.L. and Zhao, X.P., 2013. Flutemodel acoustic metamaterials with simultaneously negative bulk modulus and mass density. *Solid State Communications*, 173, pp. 14–18.

Zhang, G.M. and Harvey, D.M., 2012. Contemporary ultrasonic signal processing approaches for nondestructive evaluation of multilayered structures. *Nondestructive Testing and Evaluation*, 27(1), pp. 1–27.

Zhang, J., Hu, B. and Wang, S., 2023. Review and perspective on acoustic metamaterials: From fundamentals to applications. *Applied Physics Letters*, 123(1).

Zhao, B., Chang, B., Yuan, L. and Li, P., 2020. Influence of force load on the stability of ultrasonic longitudinal–torsional composite drilling system. *The International Journal of Advanced Manufacturing Technology*, 106, pp. 891–905.

Zhao, Z., Dai, Y., Dou, S.X. and Liang, J., 2021. Flexible nanogenerators for wearable electronic applications based on piezoelectric materials. *Materials Today Energy*, 20, p. 100690.

Zhou, T., Lan, Y., Zhang, Q., Yuan, J., Li, S. and Lu, W., 2018. A conformal driving class IV flextensional transducer. *Sensors*, 18(7), p. 2102.

Zhu, J.N., Borisov, E., Liang, X., Farber, E., Hermans, M.J.M. and Popovich, V.A., 2021. Predictive analytical modelling and experimental validation of processing maps in additive manufacturing of nitinol alloys. *Additive Manufacturing*, 38, p. 101802.

Index

Note: Page numbers in *italics* indicate figures.

Printed in the United States
by Baker & Taylor Publisher Services